OCED SCIENCE AND TECHNOLOGY INDICATORS

No 2
R&D, INVENTION AND COMPETITIVENESS

ORGANISATION FOR ECONOMIC CO-OPERATION AND DEVELOPMENT

Pursuant to article 1 of the Convention signed in Paris on 14th December, 1960, and which came into force on 30th September, 1961, the Organisation for Economic Co-operation and Development (OECD) shall promote policies designed:

- to achieve the highest sustainable economic growth and employment and a rising standard of living in Member countries, while maintaining financial stability, and thus to contribute to the development of the world economy;
- to contribute to sound economic expansion in Member as well as non-member countries in the process of economic development; and
- to contribute to the expansion of world trade on a multilateral, non-discriminatory basis in accordance with international obligations.

The Signatories of the Convention on the OECD are Austria, Belgium, Canada, Denmark, France, the Federal Republic of Germany, Greece, Iceland, Ireland, Italy, Luxembourg, the Netherlands, Norway, Portugal, Spain, Sweden, Switzerland, Turkey, the United Kingdom and the United States. The following countries acceded subsequently to this Convention (the dates are those on which the instruments of accession were deposited): Japan (28th April, 1964), Finland (28th January, 1969), Australia (7th June, 1971) and New Zealand (29th May, 1973).

The Socialist Federal Republic of Yugoslavia takes part in certain work of the OECD (agreement of 28th October, 1961).

Publié en français sous le titre :
INDICATEURS DE LA SCIENCE
ET DE LA TECHNOLOGIE OCDE
N° 2

© OECD, 1986
Application for permission to reproduce or translate
all or part of this publication should be made to:
Head of Publications Service, OECD
2, rue André-Pascal, 75775 PARIS CEDEX 16, France.

The aim of the "OECD Science and Technology Indicators" series is to make R & D data available to an ever-growing number of users and to integrate them into a much larger set of measures of science, economic activities and trade. This volume is the last to be prepared under the direction of the late Mr. Yvan Fabian who had led the Secretariat's work on science and technology indicators from its beginning in the early 1960s. The report was reviewed by the Committee for Scientific and Technological Policy and, on the Committee's recommendation, the Secretary General decided to make it generally available.

Also available

OECD SCIENCE AND TECHNOLOGY INDICATORS. Resources Devoted to R and D (February 1984)
(92 84 01 1) ISBN 92-64-12556-6   378 pages   £13.50   US$27.00   F135.00   DM60.00

THE MEASUREMENT OF SCIENTIFIC AND TECHNICAL ACTIVITIES. Proposed Standard Practice for Surveys of Research and Experimental Development. "Frascati Manual 1980" (June 1981)
(92 81 04 1) ISBN 92-64-12201-X   186 pages   £4.80   US$12.00   F48.00   DM24.00

* * *

SCIENCE AND TECHNOLOGY POLICY OUTLOOK – 1985 (June 1985)
(92 85 03 1) ISBN 92-64-12738-0   90 pages   £5.50   US$11.00   F55.00   DM24.00

EAST-WEST TECHNOLOGY TRANSFER:

- Study of Hungary 1968-1984 by Paul Marer (February 1986)
  (92 86 01 1) ISBN 92-64-12800-X   246 pages   £14.00   US$28.00   F140.00   DM62.00

- The Transfer of Western Technology to the USSR by Morris Bornstein (December 1985)
  (92 85 04 1) ISBN 92-64-12779-8   192 pages   £14.00   US$28.00   F140.00   DM62.00

- Study of Czechoslovakia by Friedrich Levcik and Jiri Skolka (August 1984)
  (92 84 03 1) ISBN 92-64-12600-7   102 pages   £4.50   US$9.00   F45.00   DM22.00

- I. Contribution to Eastern Growth: An Econometric Evaluation by Stanislaw Gomulka and Alec Nove. II. Survey of Sectoral Case Studies by George D. Holliday (June 1984)
  (92 84 02 1) ISBN 92-64-12565-5   94 pages   £6.20   US$12.50   F62.00   DM28.00

- Study of Poland 1971-1980 by Zbigniew Fallenbuchl (November 1983)
  (92 83 01 1) ISBN 92-64-12484-5   200 pages   £11.00   US$22.00   F110.00   DM49.00

TECHNOLOGY TRANSFER BETWEEN EAST AND WEST by Eugène Zaleski and Helgard Wienert (September 1980)
(92 80 02 1) ISBN 92-64-12125-0   436 pages   £22.00   US$50.00   F200.00   DM100.00

APPROPRIATE TECHNOLOGY DIRECTORY. Volume II by Nicolas Jéquier and Gérard Blanc (January 1985)
(41 84 06 1) ISBN 92-64-12643-0   394 pages   £12.00   US$24.00   F120.00   DM53.00

APPROPRIATE TECHNOLOGY DIRECTORY. Volume I by Nicolas Jéquier with the assistance of Gérard Blanc (August 1979)
(41 79 03 1) ISBN 92-64-11962-0   362 pages   £11.00   US$22.50   F90.00   DM45.00

Prices charged at the OECD Bookshop.

THE OECD CATALOGUE OF PUBLICATIONS and supplements will be sent free of charge on request addressed either to OECD Publications Service, Sales and Distribution Division, 2, rue André-Pascal, 75775 PARIS CEDEX 16, or to the OECD Sales Agent in your country.

TABLE OF CONTENTS

INTRODUCTION 7

EXECUTIVE SUMMARY 9

Part I

RESOURCES DEVOTED TO R&D

I. TRENDS IN RESEARCH AND DEVELOPMENT 17
    Trend of the Countries' Positions 17
    Research and Development Personnel 19
    Resources per Researcher 23

II. THE DECLINING ROLE OF GOVERNMENT IN RESEARCH AND DEVELOPMENT FINANCE 25
    Government's Goals in R&D Spending 29
    Who Receives Government R&D Financing? 30

III. BUSINESS: THE MAJOR R&D SECTOR 33
    Industrial R&D Compared with Value Added 34
    Industrial R&D Effort and Profitability 34
    R&D by Industry Groups 37

IV. UNIVERSITY RESEARCH AND BASIC RESEARCH 41
    Comparisons by Fields of Science 41
    National Efforts in Basic Research 43

Part II

TECHNOLOGICAL PERFORMANCE AND INDUSTRIAL COMPETITIVENESS

INTRODUCTION 47

I. TRENDS IN THE TECHNOLOGICAL SITUATIONS OF OECD COUNTRIES 48
    Patents as Measures of the Production of Technology 48
        Characteristics of the Data 48
        Overall Trends 49
        Position of the Countries as Domestic Markets 50
        Countries' Position compared for R&D and Patents 51
    Position of the Countries in the Internationalisation of Patents 52
        External Applications 52
        Impact of International Filing Procedures on Dissemination of Patents 52
    Technological Balance of Payments as a Measure of the International Dissemination of Technology 53
        Characteristics of Technological Balance of Payments Data 54
        Overall Trends 54
        Position of the Countries with Regard to the International Dissemination of Technology 55

II. TECHNOLOGICAL PERFORMANCE AND INDUSTRIAL COMPETITIVENESS 58
    Identifying High, Medium and Low R&D Intensity Industries 58
        The "High-Tec" Concept 58
        Defined in Terms of R&D Intensity 59
        Common Characteristics 60

R&D and Industrial Structures 61
    General Importance of the Three Groups of Industries at all OECD Level 61
    More about High R&D Intensity Industries 62
    R&D and Trends in Industrial Structures 65
R&D and Trends in Competitiveness 67
    Differing Trends 67
    Factors in Competitiveness 68
    Trends in Industrial Competitiveness 70

NOTES AND REFERENCES 75

TECHNICAL ANNEX
Additional Detailed Tables and Graphs 78

INTRODUCTION

This is the second OECD Science and Technology Indicators Report. Many Member governments already issue similar reports of their own which discuss a wider range of S&T indicators for their country in great depth. The interest of this OECD report is that it presents a perhaps narrower range of indicators but in an international context so that Member countries can compare their performance with those of others. Also, because OECD is an international Organisation with privileged access to certain types of data, particular attention is given to indicators of the diffusion of technology and of trade and competitiveness of highly R&D intensive industries.

The aim of the exercise is to point out the salient characteristics of, and trends in, S&T activities and competitiveness in the OECD area and to indicate the main similarities and differences between Member countries. It must, however, always be remembered that this results in a gross simplification of a very complex reality. Results for a single indicator should never be treated as a cause for national congratulation or concern, but rather as a signal that something is going on which merits further investigation. Even if several indicators are making the same kind of signal, careful interpretation is still required. For example, the purely quantitative indicator used in this report can never take fully into account the complex institutional differences between Member countries in the way their S&T systems are organised or in the size and specialisation of their economies. At international level, one of the main problems is to evaluate the role and behaviour of multinational corporations which do not fit into the "nation state" pattern of analysis on which this, as other OECD reports, is based.

Theoretically, all the indicators discussed in this report can be situated somewhere along an axis from the generation of basic knowledge via its industrial or other application to its impact on the economy and society at large. This is by no means to suggest that the process in real life is anything like as simple. There are many complex lags, loops and feed-backs. Nevertheless, the indicators can be divided into three main families measuring "inputs" into the S&T system (e.g. R&D data), "outputs" from the S&T system (e.g. patents and the technological balance of payments) and "impact" indicators (trade in R&D intensive products, productivity indices etc). Of course, none of these are complete measures; for example, R&D does not cover all the inputs into the generation of knowledge; patents, the technological balance of payments or bibliometric indicators are only very partial measures of its output and, as we shall see later, it is extremely difficult to disentangle the technological and economic factors when examining trade in R&D intensive goods.

These three broad families of indicators underlie the plan of this report. Part I deals with the most established family, R&D statistics. It covers all OECD countries and sectors of the economy and is essentially an update and extension of the main results described in the first OECD S&T Indicators Report.

Part II is more experimental and concentrates on *industrial* technology and its applications in selected Member countries. The first chapter looks at trends in the technological positions of OECD countries as revealed by patent data and the technological balance of payments and the second analyses technological performance and industrial competitiveness with special reference to trade in R&D intensive products.

The report, has been prepared and analysed on the basis of available data by the following members of the Scientific, Technological and Industrial Indicators Unit: Mr. F. Pham, general editor and co-author of Part I with Miss A. Young and Miss G. Muzart; Mr. A. Lindner and Mr. T. Hatzichronoglou, co-authors of Part II. Mrs. L. M. Grjibine and Mrs. B. Madeuf helped in preparing the first draft. The report is based on a set of working papers submitted to the 1984 meeting of the Group of National Experts on Science and Technology Indicators.

EXECUTIVE SUMMARY

OVERALL TRENDS IN R&D SPENDING IN THE OECD AREA IN THE EARLY 1980s

The recovery in R&D spending from the late 1970s

Overall, the OECD R&D expenditure picked up in the mid 1970s after stagnating in the early 1970s and increased particularly rapidly from 1979 onwards. In the 1969-1981 period as a whole, spending grew at an annual average rate of 3½ per cent, but the rate increased to 5½ per cent per year between 1979 and 1981. This upturn was particularly marked for the major countries. Results for 1983 and data for a few countries for 1984 and 1985 suggest definite slowing down.

R&D remains heavily concentrated in the largest countries

If anything, there has been a tendency for R&D expenditure to become more concentrated in the seven largest Member countries which, by 1983, accounted for 92 per cent of total R&D outlays in the OECD area.

R&D in Japan continues to grow more rapidly than in the United States or the EEC

Between 1979 and 1983, R&D grew at about twice the rate in Japan as in the United States or the EEC. Nevertheless, the United States remains the largest single R&D performer in the OECD area with just under half the total. Japan comes second but still spends substantially less than the EEC countries combined.

R&D kept pace or exceeded economic growth in most countries

The percentage of GDP devoted to R&D has generally been rising since 1979 in Member countries, especially the larger ones. Only in Switzerland and Australia has R&D fallen back in comparison with GDP plus in the United Kingdom between 1981 and 1983.

R&D has generally grown more rapidly than capital investment since 1979

This difference in growth between R&D expenditure and other more traditional forms of investment is particularly marked in the major countries and in Sweden. The trend has been accentuated by the sluggish growth in capital investment in recent years. For the major countries, Sweden and the Netherlands, R&D spending is high compared with investment in plant, machinery, equipment. The main exception is Japan which combines high R&D intensity (as a percentage of GDP) with a high rate of growth in traditional investment.

The swing from public to private support for R&D has continued so far

In the OECD area as a whole, there has been a marked shift in the source of R&D funds, with the Business sector supplanting Government as the single largest source of funds in 1979. Business R&D finance also grew more rapidly than government funds in most individual countries. The major recovery on defence R&D in the United States has not succceded in reversing this trend for the OECD area as a whole as the business share continued to grow in 1983.

TRENDS IN PUBLIC R&D FUNDING

Government R&D finance is growing in absolute terms and holding its share of national budgets

Overall, governments financed about 45 per cent of total OECD R&D in 1983. Funding has been growing at fixed prices since 1979 in all countries except Switzerland, Norway and Portugal. In the large countries at least, it has increased more rapidly than total government expenditure, after a period of relative decline during the 1970s.

Japan and the United States are at the opposite end of the spectrum in terms of the destination of government R&D funds

In the United States, industry is the biggest beneficiary, receiving between two-fifths and one-half of Federal support as against one-quarter each for universities and government-owned research establishments. In Japan, industry receives only 5 per cent (about the same as the PNP sector) with the balance of 90 per cent divided about evenly between government laboratories and the universities.

Overall, most OECD government R&D funding goes to defence and space programmes and these are growing, bringing new contracts to industry

Half of all government-financed R&D in the OECD area goes to defence and space programmes but that largely reflects the weight of the United States in the aggregate figures. Two other countries – France and the United Kingdom – spend heavily on defence and space research. Such programmes are also relatively important in Germany and Sweden. Much of the funds are paid to industry, and the recovery in defence R&D funding, which has been especially marked in the United States and Sweden, may reverse the trend towards declining government support for industrial R&D in these countries.

Energy R&D funding is dropping back after a period of rapid growth

Most governments are scheduling little growth or even a decline in their support for energy R&D in 1980s. The decline is particularly marked in the United States, Belgium and Norway.

Most governments continue to commit a high share of funds for the "advancement of knowledge"

Most Member governments spend more money for general non-oriented support for R&D (including university research) than they do for R&D for any other individual policy objective such as energy, industrial development or defence (the No. 1 for the OECD area as a whole).

UNIVERSITY RESEARCH

The Higher Education sector accounts for a declining share of national R&D efforts despite a slight recovery in spending

Although the general upturn in R&D spending in the OECD area since 1979 affected all sectors of the economy, the Higher Education sector was the least favoured and the steady decline in its share of total R&D continued. This was particularly marked in Canada, Norway and Japan.

The United States is the clear leader for university R&D in the natural sciences and Japan in the social sciences and humanities

American universities are responsible for over half the OECD total. Although Japanese universities spend at least twice as much on R&D overall as their counterparts in the three major European countries, they perform less R&D in the natural sciences. Even allowing for the fact that it is particularly difficult to measure academic R&D in the social sciences and the humanities, it is clear that Japanese universities are the major performers in the OECD area, responsible for one-third of the total.

Universities' spending on R&D in engineering and in the medical sciences is broadly evenly divided between the United States and the EEC

Nevertheless, the United States leads for university engineering research and the EEC for academic medical research.

BASIC RESEARCH, APPLIED RESEARCH AND EXPERIMENTAL DEVELOPMENT

Basic research got about 15 per cent of the R&D funds in the OECD area in 1981

On average, two-thirds of basic research was done in the Higher Education sector with a slight contribution from the PNP sector. The remainder is carried out equally in industrial and government laboratories.

Research versus development

Although the United States is the major spender for all types of R&D activity, its role is particularly strong in experimental development whereas the EEC is better placed for research. Japan falls between these two patterns.

Basic research in industry and in the public sector

Almost half of the basic research in government laboratories in the OECD is performed in EEC countries. The United States and Japan, on the other hand, are better placed for basic research carried out by industry, together performing two-thirds of the OECD total.

R&D PERFORMED BY INDUSTRY

Industry is increasing its role in OECD R&D efforts

The share of all R&D in the OECD area carried out by industry has continued to rise, reaching almost two-thirds of the total in 1983. Growth since 1979 has been particularly marked in Japan. Industrial R&D in the United States increased at the average rate for the OECD area, whereas in the EEC it did less well because of slow growth in general and an actual decline in volume in the United Kingdom between 1981 and 1983.

Business funds were mainly responsible for increased industrial R&D

In most countries, notably Japan, growth in industrial R&D since 1979 has been financed from business sources though in a few, notably France, Italy, the Netherlands, Ireland, Spain and Canada, government support has also played a part.

In the other countries, government finance followed the trend for business funds. However, government finance for industrial R&D in the United States has strengthened since 1981.

R&D represents a growing share of value added in the Business Enterprise sector, especially in manufacturing

The percentage of industries' resources going to "intellectual" investment seems to be rising. The average for the OECD area is 1¾ per cent of value added in 1983 but the intensities tend to be higher in large countries (plus Sweden and the Netherlands) than in small ones, and in manufacturing than in other industries.

The degree to which company-financed R&D weighs on industries' "earnings" varies considerably between countries

Such R&D spending corresponds to between 3 and 25 per cent of gross operating surplus in manufacturing industry and is growing in all Member countries. This ratio depends partly on the level and type of R&D undertaken but also on industry's ability to generate profits. Compared with the major other countries, Japanese manufacturing industry pays a lower share of value added in wages and salaries and thus has a greater surplus available for all kinds of investment.

R&D in the electrical/electronic industries and the machinery industries (including computers) has continued to grow more rapidly than in other groups. Aerospace R&D is picking up

Although changes in the balance of R&D efforts between groups of industry at all-OECD level take place only slowly, electrical/electronics and machinery (including computers) continue to pull ahead. R&D in the transport equipment industry (excluding aerospace) has slowed down after a long period of growing faster than the average for industry. Conversely, R&D in the aerospace industry which was falling back prior to 1979, showed new vigourous growth until 1982. R&D in the chemical industries continues to grow at below average rates.

Japan and the United States have strengthened their R&D positions in different industries whereas the EEC share has dropped for almost all

Japan continues to perform a growing share of all industrial R&D especially in the electrical/electronics and transport equipment industries. The United States has reaffirmed its predominant role in aerospace (now in a hesitant recovery) and has also improved its position in the machinery group, thanks to R&D on computers and scientific instruments. A slow decline is evident in the EEC share of total OECD R&D in virtually all industry groups.

TRENDS IN THE TECHNOLOGICAL SITUATIONS OF OECD COUNTRIES

Patenting activities generally declined in the OECD area from 1965 to 1978 but recovered thereafter

While not a very accurate measure, patent data do give some insight into a country's position as a producer of technology and as a source of internationally disseminated technology. National patent applications declined in OECD countries during the early and mid-seventies, except in the United States, Japan and one or two smaller countries. This downturn was mainly determined by foreign applications. This trend has, however, been reversed since 1978 if international patent channels are taken into account.

Domestic patenting has lagged behind R&D

The "apparent productivity" of research in the OECD area seems to be declining in that the "output" of domestic patent applications per researcher has been falling in most countries. It is not clear whether this reflects slower generation of knowledge per R&D input or a decline in the propensity to patent R&D results.

The United States and Japan are the largest and most dynamic patent markets

The dynamism of the Japanese and United States' patent markets has different origins: the particularly strong increase in patenting in Japan was primarily fuelled by Japanese inventors whereas foreign inventors played that role in the United States where the proportion of domestic patent applications has fallen to under 60 per cent. Japan has become the main market for patents, measured in terms of applications, accounting for over 33 per cent of the OECD total in 1983, followed by the United States (14 per cent). However, it should be borne in mind that Japanese patents are single-claim patents and thus lead to exaggerated international shares and that the above situation is reversed for patents actually granted, with the United States clearly in first place followed by Japan.

Japan patents more than other OECD countries and does not follow the general decline in patent productivity

Relating domestic patent applications to R&D confirms the Japanese high propensity to patent. In 1981 Japan spent at least a sixth of total OECD business enterprise R&D but generated half of all domestic patent applications. The reverse was true for the United States. The EEC and other countries had proportional shares. Furthermore, Japanese researchers continue to generate as many patents per head as in the 1970s.

Japan has also stepped up its patenting abroad though it is still well behind the United States

The United States is responsible for about one-third of all external patent applications in the OECD area, down slightly over the decade. Whereas the German share has remained stable over time but fell back in 1983, Japanese external patent applications have shown six-fold growth from a low base, reflecting its successful penetration of foreign, high technology markets.

International filing grew substantially from 1978 following the introduction of new procedures

The main feature of recent years has been the increase in transborder applications, mainly of patents filed through new international channels. Non-signatory countries to the European Patent Convention (EPC) took advantage of these channels to enlarge the area of protection of their patents abroad without experiencing an increase in foreign patents. Smaller signatory countries were more affected by this opening to foreign technology than larger ones.

Technology transfer is highly concentrated in the OECD with the United States being by far the biggest exporter according to Technological Balance of Payments data

The United States remains far and away the biggest exporter of technology in the OECD, with a major share of total receipts and a very large surplus in its technological balance of payments. On a much smaller scale, the United Kingdom is the only other major OECD country to have a technological payments surplus.

Although still showing a deficit, Japan has become less dependent on imported technology and now plays a more important part in the international dissemination of new technology

Japan's receipts have increased at fixed prices, as have France's but, unlike France, its payments have fallen (as have those of the United States and the United Kingdom). In absolute terms, Japan earns about as much as the United Kingdom from technology sold abroad, twice as much as Germany.

International technology transfers play a minor role in major countries compared to their own R&D efforts

For the five major OECD countries, plus Sweden and Australia, international purchases and sales of technology correspond to less than one-fifth of their business enterprise R&D efforts, whereas this ratio is significantly higher the smaller and/or less industrialised the country. Thus, Spain and Portugal spend one-and-a-half times as much on paying for foreign technology as for their national industrial R&D effort.

Technology transfer is largely determined by the strategies of multinationals

Statistical evidence from a few countries suggests that payments made by subsidiaries of foreign companies can be largely responsible for technological deficits. The relationship between technology transfer and direct investment abroad is obvious and merits further investigation.

TECHNOLOGICAL PERFORMANCE AND INDUSTRIAL COMPETITIVENESS

High R&D intensity industries grew the fastest at all-OECD level between 1970-1983 measured in terms of production, internal demand, imports and exports

The recession following the two petroleum shocks practically affected only medium and low R&D intensity industries. The high R&D intensity industries (where R&D is equivalant to 11 per cent of turnover on average) grew rapidly, headed by the electronics industry, computers and scientific instruments. Production was clearly up in the pharmaceutical industry but there was only moderate growth in trade (partial substitution of trade by direct investment). Nevertheless, the high R&D intensity industries are still responsible for only a very small share of manufacturing production (11 per cent on average) and only a slightly higher one of trade (16 per cent on average).

High R&D intensity industries are responsible for about half of all R&D expenditure in manufacturing industry and get between 60-80 per cent of direct government R&D grants and contracts received by manufacturing industry

Government R&D support is particularly heavily concentrated in highly R&D intensive industries (over 40 per cent) in the United Kingdom, France and the United States where the aerospace industry plays a particularly important role. Japan is an exception in that the relatively little direct government finance for manufacturing R&D (about 2 per cent of the total) is mainly destined for low R&D intensity industries (60 per cent).

Although employment is growing in high R&D intensity industries, the effect is small compared with job losses in the other groups

During this phase of adjustment, it was the electronic industry which created jobs the most rapidly. However, except in Japan, this growth in employment was insufficient to offset the number of jobs lost in medium and low R&D intensity industries which were still responsible for 80-85 per cent of manufacturing employment in OECD countries in 1982.

The United States is the only country which has a positive overall balance earned exclusively by high R&D intensity industries up to 1983

All the other large countries have absolute balance earned largely by medium R&D intensity industries (motor vehicles, chemicals, non-electrical machinery) whereas in the United States highly R&D intensive industries are mainly responsible. The aerospace and computer industries alone contribute 56 per cent of their surplus to the manufacturing account.

Japan is the only OECD country with a positive trade balance for all three categories of industry (high, medium and low R&D intensity)

The Japanese pattern of specialisation is the best adapted to foreign trade. Especially since the first petrol shock, Japan has committed itself to fast growing industries (electronics, computers, scientific instruments, machinery, motor vehicles) and has progressively given up its "niches" in mature industries (textiles, footwear, leather, tobacco, furniture, non ferrous metals). Nevertheless, no type of industry has been wholly abandoned, hence its spectacular surpluses for all three categories.

United States competitiveness has declined 1970-84 even in high R&D intensity industries (loss of export markets, increased import penetration)

Export markets have been lost by the aerospace, telecommunications, consumer electronics, motor vehicle and steel industries. The decline in the trade balance for high R&D intensity industries occurred because, although production grew at the same rate as domestic demand, exports grew twice as fast and imports three times more rapidly. In consequence, import penetration tripled over ten years.

EEC competitiveness declined for the high R&D intensity industries and rose for the low intensity ones

Two-thirds of the decline in competitiveness in the

highly R&D intensive group could be attributed to the electronics and computer industries. The competitiveness of medium R&D intensity industries was generally stable despite the weakening of the United Kingdom's motor vehicle industry. The gain in export shares by low R&D intensity industries was mainly due to the food and drink industries and to a lesser extent the wood and petroleum industries.

The EEC is still the zone which is the most exposed to international competition for all three categories of manufacturing industry

Japanese high R&D intensity products compete mainly on foreign markets whereas American goods mainly face competition in home markets. EEC high R&D intensity industries are twice as exposed on home and on foreign markets as those of either the United States or Japan.

Part I

RESOURCES DEVOTED TO R&D

Chapter I

TRENDS IN RESEARCH AND DEVELOPMENT

The OECD countries spent 192 billion on R&D[1] in 1983[2]. This was up 50 per cent in volume[3] over the last twelve years, giving an average annual growth rate for the OECD area of 3.5 per cent. Growth was, however, uneven: rates were low until 1975 and were slower than previously. They subsequently picked up, reaching nearly 5 per cent for the first three years of the 1980s.

Provisional data for a significant number of countries suggest that at best overall growth is continuing at the same pace in 1984 and probably 1985 (though, it should be remembered, forecasts generally tend to be over optimistic).

Trend in the countries' positions

Following, are a few major trends in the breakdown of R&D expenditure in the OECD area:

i) This expenditure is still very highly concentrated in the main countries and has tended to become even more so. Taken together, the five largest R&D spenders were responsible for 86 per cent of the OECD total in 1981. The share of the top seven was 91 per cent.

ii) There are striking changes in position which reflect disparities in the rate of growth of R&D resources in various geographic areas:

Table 1.1
Percentages by main OECD area[1]
When total OECD = 100

	R&D expenditure				Researchers			
	1969	1975	1981	1983	1969	1975	1981	1983
Other	7.5	8.2	7.9	7.8	8.2	8.7	8.8	8.9
EEC	28.0	30.8	29.6	28.7	25.2	26.5	25.4	25.3
Japan	9.3	13.5	16.1	17.4	18.9	24.3	24.0	24.8
United States	35.1	47.5	46.3	46.0	47.8	40.4	41.8	41.0

1. For more details see Appendix, Table 1 and 2.
Source: OECD/STIIU Data Bank, November 1985.

- The uptrend in Japan whose R&D expenditure in 1983, as officially reported, amounted to over 17 per cent of the total for the area (as compared with 11 per cent in 1971), a doubling of expenditure in ten years and an average growth rate over twice that of the OECD area as a whole.
- A reduction in the prevalent role of the United States where research and development spending had already reached a very high level by 1969: 55 per cent of the OECD total. It levelled off during the first half of the 1970s, but then returned to an average annual growth of over 4 per cent from 1975 to 1983. Nevertheless, because of the impressive rise in Japanese expenditure, the United States' share in total R&D expenditures in the OECD area dropped to 46 per cent in 1983.
- The stability of the European Community expenditures during the period, which accounted for almost one-third of all investments in R&D. Although its R&D grew at the same pace as the OECD average, there were substantial differences between the ten countries. Germany, and to a lesser extent Italy and France, were the motors of growth, while the United

Graph 1
R&D resources in the OECD area

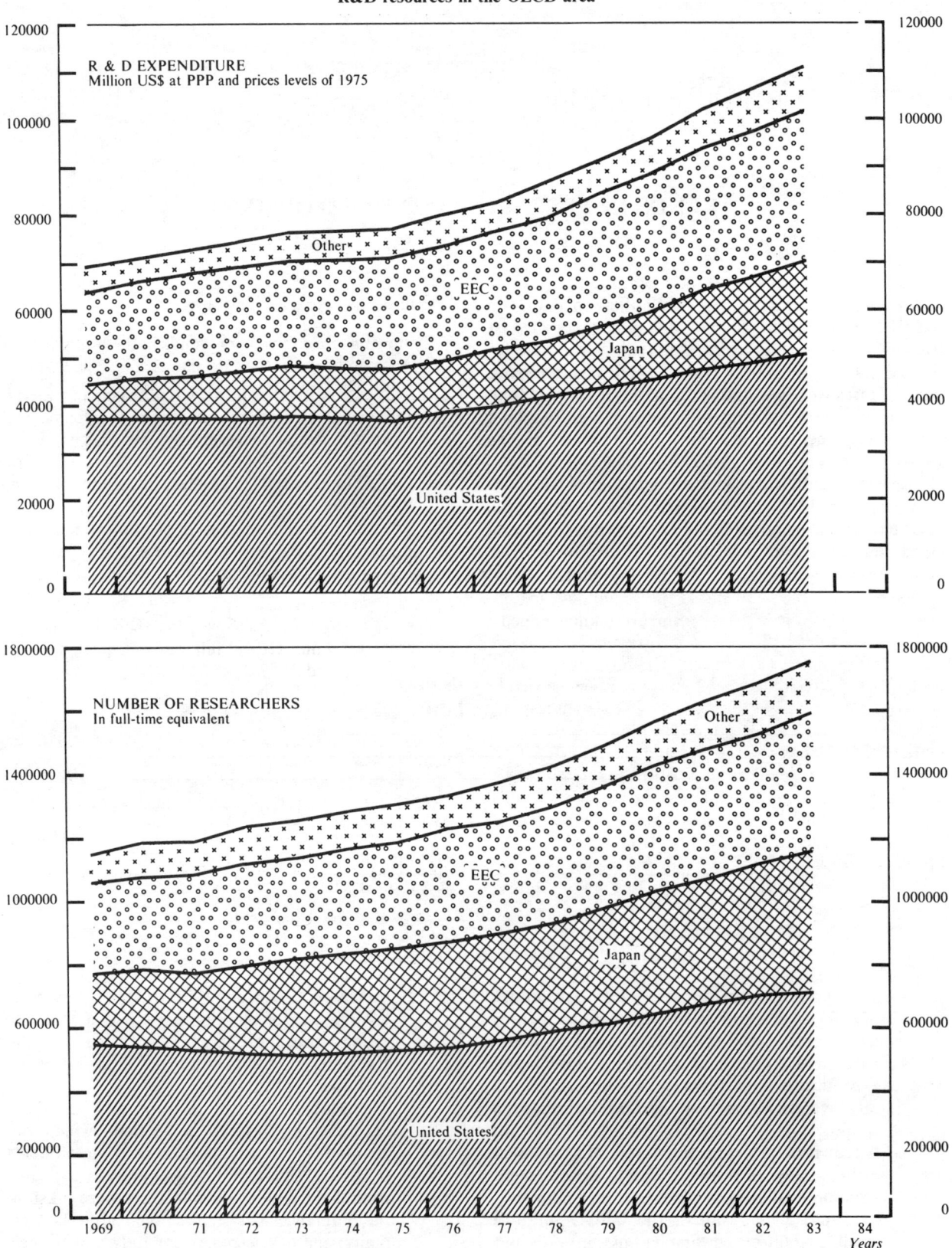

Source: OECD/STIIU Data Bank, November 1985.

Kingdom and the Netherlands acted as brakes.
- The remaining countries of the OECD area, which represent one-half of the membership, invested only around 8 per cent of the R&D total. Their share remained broadly stable throughout the period under consideration.

Research and National Wealth

The advantage of the preceding comparisons of absolute amounts of money spent on research and development is that they give a clear idea of the actual situation. However, one cannot expect a small country to invest the same absolute amount in R&D as a large country. That is why it is useful to resort to intensity indicators, which reflect the levels of effort according to the size of a country's economy or its population. One of the most commonly used of these intensity indicators is gross domestic expenditure on research and development (GERD) as a percentage of gross domestic product (GDP) (Graph 2).

This indicator, calculated for the OECD area as a whole, has risen steadily since 1979, reaching roughly 2.2 per cent of GDP in 1983. This very high level was due primarily to the larger countries as the median was only 1.3 per cent.

The preponderant role of the United States, Germany, Japan, France and the United Kingdom in the R&D potential of the West is commonly acknowledged. But their role could become even greater because of economic difficulties, (R&D, it is felt, can be a key way to overcome these difficulties) and because of the fierce competition in international trade and the likely development of armaments. It is highly unlikely that the relationship between the top five and the other OECD countries will change in the near future, especially since the top five have, in addition to their size, a higher R&D intensity: their R&D/GDP ratio is almost twice that of the other OECD countries. Judging from current R&D data and from national forecasts, plans and projections, and allowing for growth, R&D in the United States, Japan and Germany may be expected to reach 3 per cent of GDP by the end of the eighties whereas in the majority of the other OECD countries the maximum will be 1.5 per cent.

The GERD/GDP ratio has risen in almost all OECD countries since 1979/80, usually quite rapidly. However, except in Japan and one or two small countries where there was a substantial real increase in R&D expenditures, this largely reflects the semi-stagnant state of national economies. For example, all OECD GDP was up only 4 per cent in volume between 1980 and 1983 and EEC GDP was up only 1 per cent over the three years. It is worth noting that, according to official data, Japan overtook Germany in terms of R&D as a percentage of GDP for the first time in 1983, putting it in second place close behind the United States (see Graph 2).

R&D Investment and Traditional Investment

The preceding observations on the differences in R&D intensity in relation to national wealth (GERD/GDP) between large and small countries as well as on the evolution of national intensity indicators are even more striking as far as the GERD/GFCF ratio is concerned. (This ratio compares R&D investments with traditional investments in equipment, machinery and construction.)

Indeed, the difference between the larger and smaller countries becomes even clearer when R&D expenditures are compared with traditional investments. As a general rule, the smaller countries invest less proportionately than the large countries in research and development, but devote a higher percentage of their resources to traditional investments.

When gross fixed capital formation is related to gross domestic product, it appears that smaller countries spend the same or an even higher proportion of GDP on traditional investment (24 per cent in the five smallest countries as against 20 per cent in the five largest ones). Thus it follows that the gap between the smaller and larger countries widens if one compares gross expenditures on research and development with GFCF; the average figure for the five largest countries in the OECD area is 10 per cent, as against only 3 per cent for the five smallest countries.

The GERD/GFCF ratio soared over the past few years as traditional investment levels were held down and essential capital projects were postponed because of the recession. Thus GFCF fell 1 per cent in volume between 1980 and 1983 in the OECD as a whole and 6 per cent in the EEC.

The case of Japan warrants further explanation. A first glance at Graph 3 ratios suggests that Japan shares an investment pattern with the medium-sized countries rather than the other large industrialised ones. However, what this does not show is that the Japanese ratio appears rather low only because of exceptionally high conventional investment. For example, GFCF represented 28 per cent of GDP in 1983 in Japan (as against 20 per cent in the five major countries combined) or three-quarters of EEC capital investment.

As we shall see in the second part of this report, the interaction of these two types of investment (R&D and conventional) apparently helped the Japanese economy absorb technical innovations very quickly. This is corroborated by the remarkable upswing of Japan's trade balance in R&D intensive industries compared with those of the other large OECD countries. An economy that is expanding more rapidly than others is able to renew its plant faster and, consequently, to incorporate technological advances sooner.

Research and Development Personnel

Curiously enough, it is more difficult to obtain up-to-date information on R&D personnel for the

Graph 2
Intensity of research and development investments*

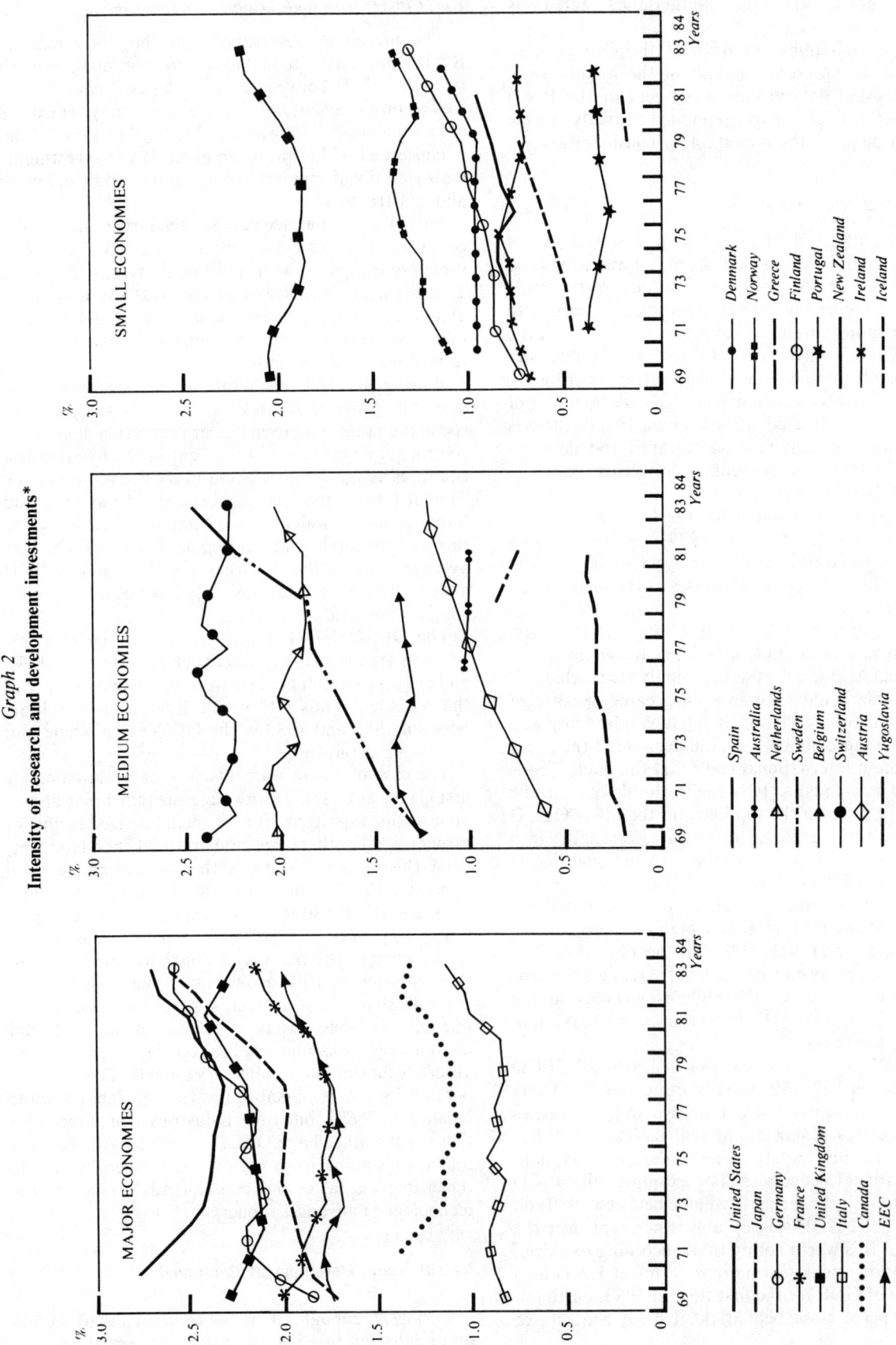

* Total R&D expenditure (all fields of sciences) as a percentage of Gross Domestic Product.
Source: OECD/STIIU Data Bank, November 1985.

Graph 3
Research and development related to traditional investments*

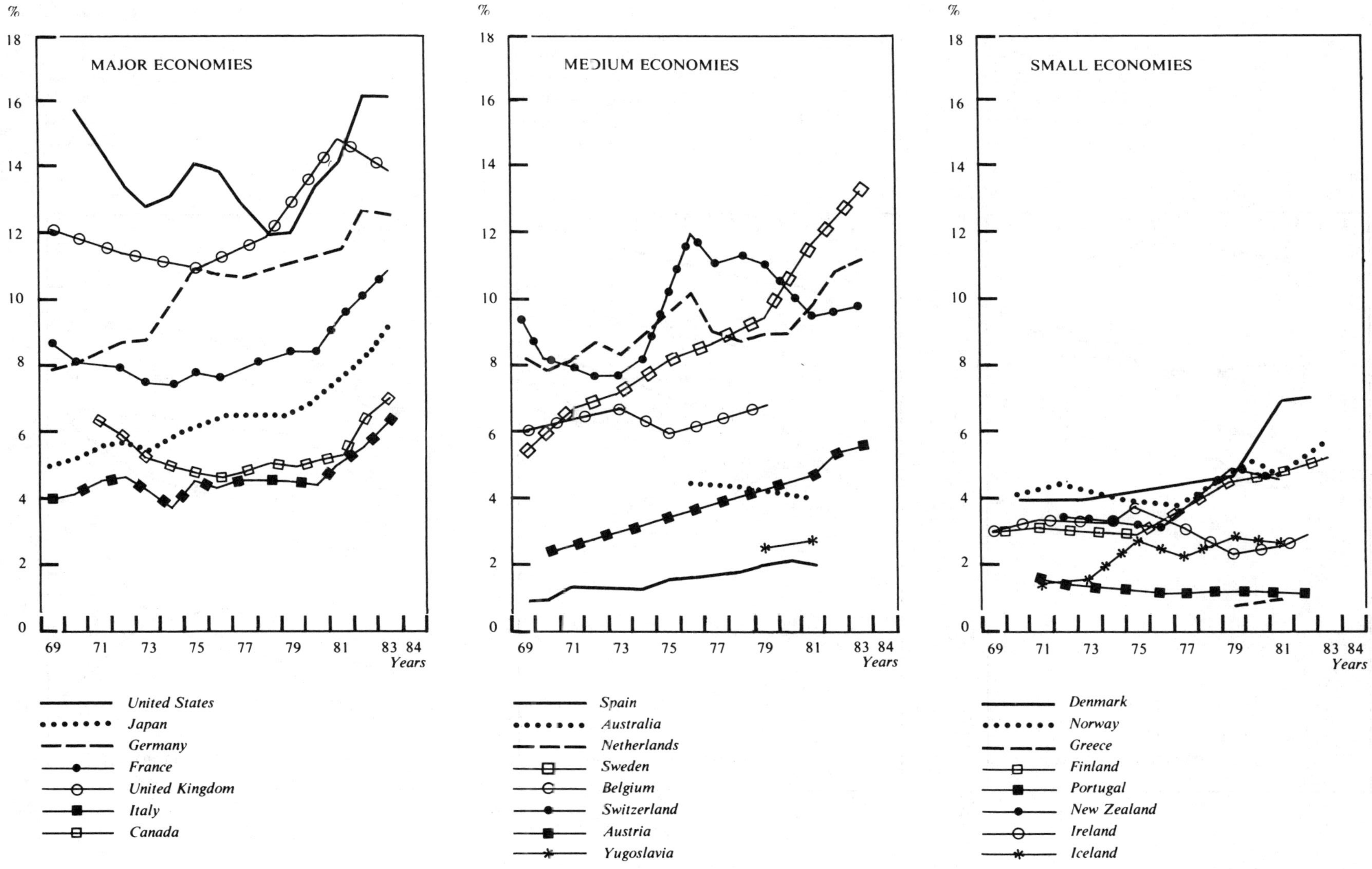

* GERD as a percentage of Gross Fixed Capital Formation.
Source: OECD/STIIU Data Bank, November 1985.

Graph 4
Number of researchers in relation to total labour force
Per thousand

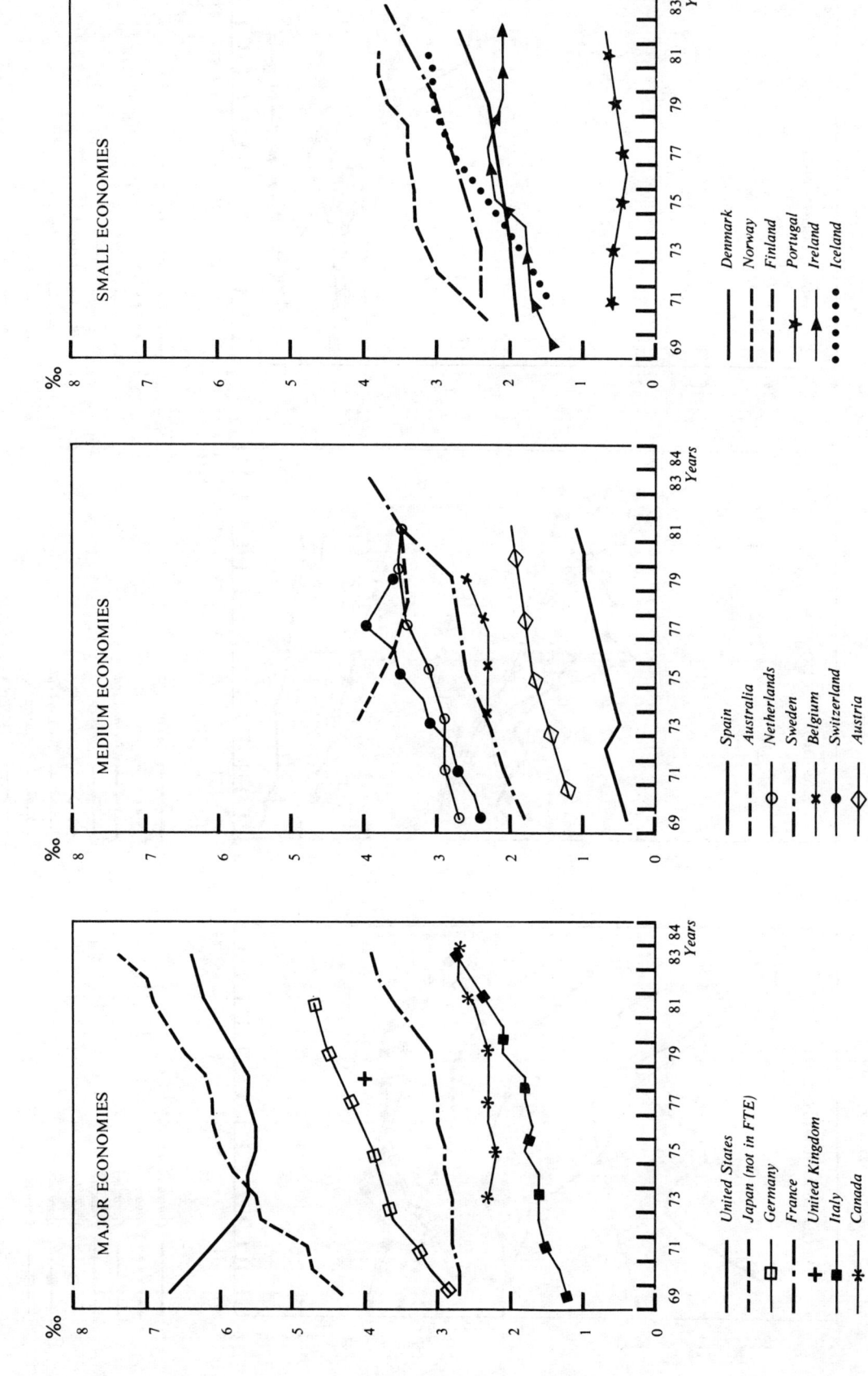

Source: OECD/STIIU Data Bank, November 1985.

OECD area than for expenditures. At the time of writing, it was possible to obtain almost a full set of R&D expenditure data for 1983 from early returns to the biennial OECD survey, national sources, etc. whereas similar personnel figures could be found for only about one-quarter of the countries concerned. In consequence, international comparisons have had to be made using 1981 data. A quick check of countries where both 1981 and 1983 data are available suggests that there has generally been little change in the ranking and structure of R&D personnel over the two years.

In 1981, the total number of researchers employed in the OECD area stood at approximately 1.65 million expressed in full-time equivalence (except in Japan). If the Japanese statistics are adjusted to obtain the equivalent in full-time personnel, using the international norms proposed by the Frascati Manual[4], the OECD total is 1.55 million. From 1969 to 1981, the number increased on average by 3 per cent a year but, as was the case for R&D expenditures, growth was more rapid (4.5 per cent) after 1979. The pattern of distribution of researchers between countries differs significantly from that for R&D spending. Countries where the "non-competitive" Government and Higher Education sector perform above average shares of the national R&D effort, for example Italy, Spain, Australia, Ireland and Japan, generally rank higher on the international scale in terms of R&D personnel than for R&D spending.

The share of the United States in the total number of researchers employed in the OECD area picked up quite sharply during the second half of the 1970s, after a rather long period of decline. The most recent available statistics seem to indicate that the average annual growth consolidated at around 3 per cent a year from 1980 to 1984.

There has also been a notable increase in the number of researchers since 1980 in France, Sweden, Italy and Portugal. In contrast, growth rates have slowed down in Germany, the Netherlands, Spain, Norway and Ireland, in some cases after substantial increases in the 1970s.

As for Japan, its share of researchers has remained basically stable since 1975. As was indicated above, official Japanese statistics are inflated by OECD evaluation standards, and especially as far as the number of researchers is concerned, because of the impact of the Higher Education sector where, in Japan, any member of the teaching staff who "regularly" performs research and development as part of his university activities is counted as a full-time researcher. If Japanese data are adjusted to make them more comparable with those of other countries, the total number of researchers in Japan comes to around 75 per cent (instead of 95 per cent) of the EEC total and 45 per cent (instead of 57 per cent) of that of the United States.

The same trend, revealing growing commitment to research and development, can be observed if one compares the number of researchers with the total labour force. Between 1970 and 1981, the ratio increased steadily in practically all the OECD countries, with the exception of Australia and the United States, where the ratio began to pick up only in 1977-78. This steady rise was helped by the generally modest growth in the labour force (on average, 1.2 per cent a year for the OECD area).

By 1981, roughly 50 in every 10 000 actual or potential workers in the OECD area were engaged as researchers. This ratio is far higher in the United States and Japan (adjusted figures) than in the major European countries. The average ratio for the EEC remained relatively low, held down primarily by the smaller Member states which, as we saw earlier, are less R&D intensive than the larger countries.

Resources per Researcher

There has apparently been no improvement in the amount of material resources or the number of assistants available to researchers, either at the national or sectoral level. This state of affairs has given rise to very different interpretations, reflecting the varying views of the researchers themselves, or of the government departments and universities employing them which are proposing "structural" reforms. These explanations have, however, become more nuanced with time.

Spending per researcher seems to be only modestly affected by size of country. According to the data available, the average for the OECD area in 1981 was about $100 000 per researcher (including support staff and equipment) and the median $70 000 as against $110 000 for the five largest countries and $60 000 for the five smallest ones.

Despite this, and allowing for the difficulties of making international comparisons, one can say that the average cost of a researcher seems quite low in certain countries. This is the case of Japan compared both with the EEC average and the average for "other" countries. The latter is pushed up by the presence of Sweden, Switzerland, Canada and Austria which have some of the highest expenditures per researcher in the OECD area. In Australia and Spain, on the contrary, R&D spending per researcher is under two-thirds of the OECD average.

In most countries there is no significant variation in R&D spending per researcher between the Industry, Government (and the Private Non-Profit) sector, and little sign of the expected higher figure for industry. However, spending per researcher is almost always significantly lower for the Higher Education sector than in the other sectors: the industry/university ratio in the OECD area is generally between two and one: in certain countries, such as Spain, Belgium, Italy, Ireland and Switzerland, the ratio even exceeds four. These differences call attention to the rather modest level of resources available to each researcher in this sector, a level that has hardly improved since 1970 for the OECD area as a whole as well as for the majority of Member

Table 1.2
Changes in the level of R&D resources per researcher

	Expenditure per researcher, 1981							
	Business enterprise		Higher education		Government + PNP		All sectors	
	$1 000	1971=100	$1 000	1971=100	$1 000	1971=100	$1 000	1971=100
United States	104	100	108	98	130	91	108	96
Japan	80	122	38	115	106	164	65	126
EEC[1]	149	109	52	81	112	127	98	106
Other[1]	117	99	54	97	82	111	84	102
Total[1]	113	104	63	92	107	109	89	102

	Supporting staff per researcher, 1981							
	Business enterprise		Higher education		Government + PNP		All sectors	
	Ratio	1971=100	Ratio	1971=100	Ratio	1971=100	Ratio	1971=100
United States	n.a.	n.a.	n.a.	n.a.	n.a.	n.a.	n.a.	n.a.
Japan	0.90	63	0.31	55	0.95	83	0.65	64
EEC[1]	2.54	78	0.69	68	1.90	83	1.59	76
Other[1]	2.00	78	0.76	82	1.60	108	1.45	86
Total[1]	1.80	71	0.59	66	1.48	89	1.23	73

1. *EEC* excluding the United Kingdom and Greece.
   Other, excluding New Zealand, Turkey (and Spain for the supporting staff ratio).
   The *total* includes countries for which data are available.

countries. In one-third of all countries (regardless of size) approximately $30 000 a year was available for all of his or her needs (including equipment, administrative and salary expenditures).

The general stagnation of total expenditures on research and development per researcher has, in almost every case, affected capital expenditure more adversely than operating expenditures. The latter take longer to cut, primarily because of salaries. It is worth noting that in Japan the level of capital expenditure per researcher is comparable to that of Germany or France, although the total cost is much lower, as noted earlier.

In addition to material and equipment, researchers have to be able to call upon a team of supporting staff, such as technicians, semi-skilled workers and administrative personnel. Here, too, the situation has deteriorated. For every ten researchers, there were about 17 people assisting them in 1971 but only 12 by 1981 for those countries for which data were available.

In short, these are the principal observations that can be made concerning the resources available to R&D researchers, those in the industrial sector being relatively better off than in the state-run sector or in universities. As a concluding remark, one should add that, unless the trend for the hiring and replacement of researchers picks up (3 per cent on average between 1969 and 1981 for the OECD area), it is likely that researchers will tend to be older as a group. Added to the resource trends identified above (a relative drop in the amount of supporting staff and equipment), this could have very negative effects on research work. This state of affairs was recently brought to our attention by Dr. Masaru Goto, Director General of the National Research Institute for Non-Organic Matter at Tsukuba, who remarked that he had over a hundred researchers working in his institute. The quality of their work and their productivity were widely recognised as being well above the national average. However, their average age was 41, and replacement would be a problem. Yet, given their age, it would be increasingly difficult to ask our researchers to continue doing tasks that were not really part of their research work and which should ordinarily be given to technicians and maintenance personnel to do.

Chapter II

THE DECLINING ROLE OF GOVERNMENT IN RESEARCH AND DEVELOPMENT FINANCE

The most striking change in R&D funding over the past decade, the shift in balance between government and business support, has continued into the 1980s. Together, they still account for the major portion of R&D funds but, since 1979, business has supplanted government as the single largest source of funds in the OECD area as a whole.

Table 1.3

Percentages by main OECD areas[1]

When total OECD = 100

	Government finance				Business finance			
	1969	1975	1981	1983	1969	1975	1981	1983
Other	6.9	8.3	8.5	8.6	7.7	7.9	7.2	7.1
EEC	26.3	31.2	30.9	30.8	30.2	30.2	28.3	27.1
Japan	5.2	8.0	9.7	9.5	13.2	17.0	19.9	22.1
United States	61.6	52.4	50.9	51.1	49.0	44.9	44.6	43.7

1. For more details see Appendix, Tables 3 and 4.
Source: OECD/STIIU Data Bank, November 1985.

This is clearly illustrated by Graph 5 which shows public and business financed R&D expenditures by geographic zones. The new balance is due primarily to developments in the United States and Japan as, in the European Community, business support grew only slightly more rapidly than government finance.

In a country-by-country comparison, it can be seen that this shift generally became pronounced beginning in the mid-1970s. Graph 6 plots this trend, giving the ratio of private to public R&D funds; the relative steepness of the curves indicates the magnitude of the change. The only notable exceptions to the general trend were Switzerland, Italy and the Netherlands as the recent apparent drop in the share of business funds in Sweden, France and Spain was due partly to upward technical adjustments to government-finance data. However, there are signs that the balance may be swinging back slightly in the United States.

In absolute amount, total government support for R&D in the OECD area came to about $85 billion in 1983. Between 1979 and 1983 it should be up by about one-sixth with rather faster growth in the European Community than in the other countries combined.

Looking at the individual countries (see Annex, Table 3), it is the United States that accounted for the largest share – 50 per cent – of all government-financed R&D in the OECD area in 1981, and this despite a steady relative decline. (As we saw earlier, the United States also holds first place for gross expenditure on R&D, though it now accounts for less than half the total.) In absolute terms, the United States' government spending was up only just over 10 per cent between 1979 and 1983 but is expected to accelerate in 1984 and 1985 to finish about one-third higher than in 1979. Japan, which from a national expenditure standpoint ranks as an R&D giant, occupies a more "normal" position as regards government funding which is equivalent only to that of Germany or to one-fifth that of the United States. If one were to adjust the official data series generally used which, for technical reasons, appear to

Graph 5
Comparison of public and private R&D funding

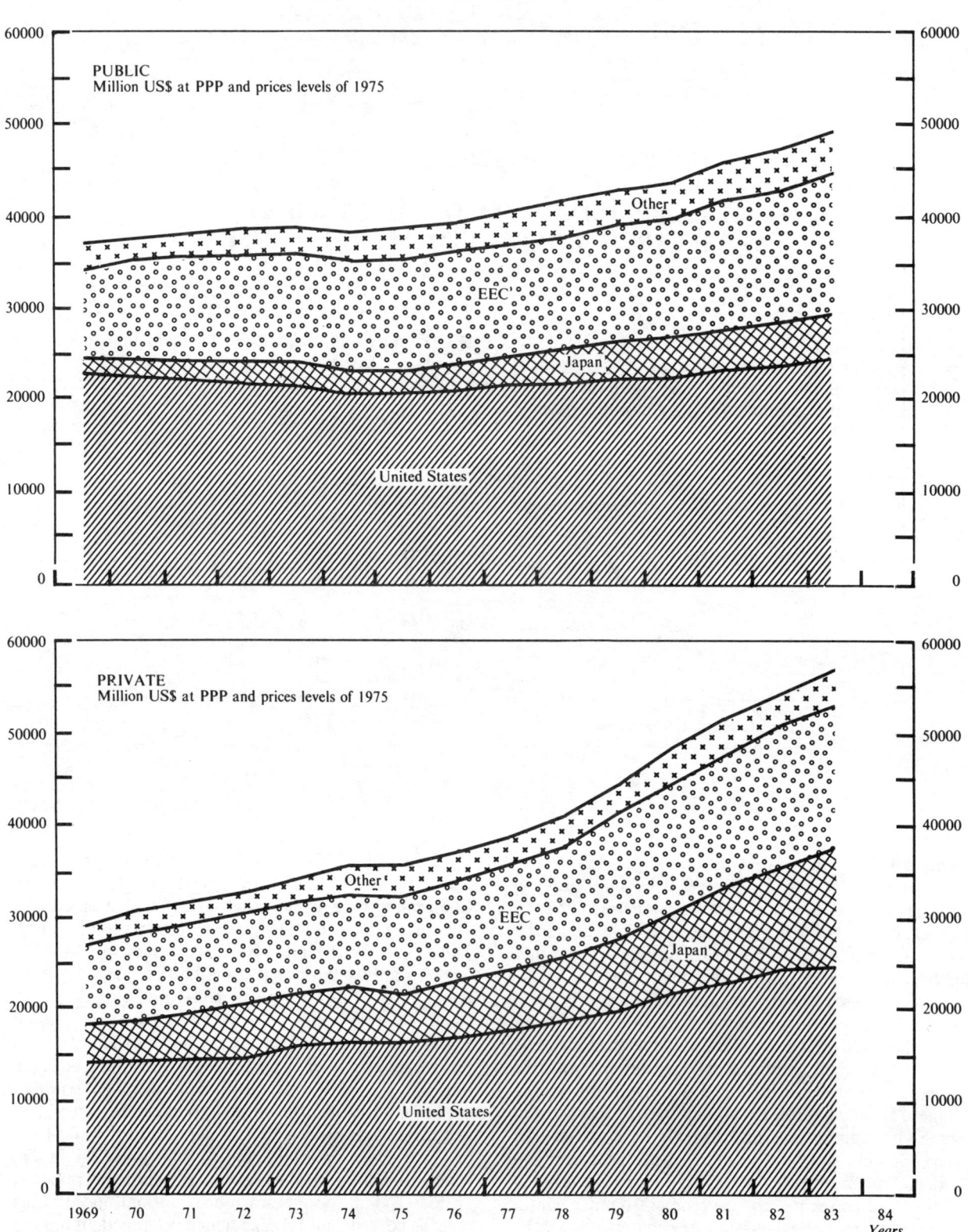

Source: OECD/STIIU Data Bank, November 1985.

Graph 6
Ratio of business to public GERD funding
When public GERD funding = 100

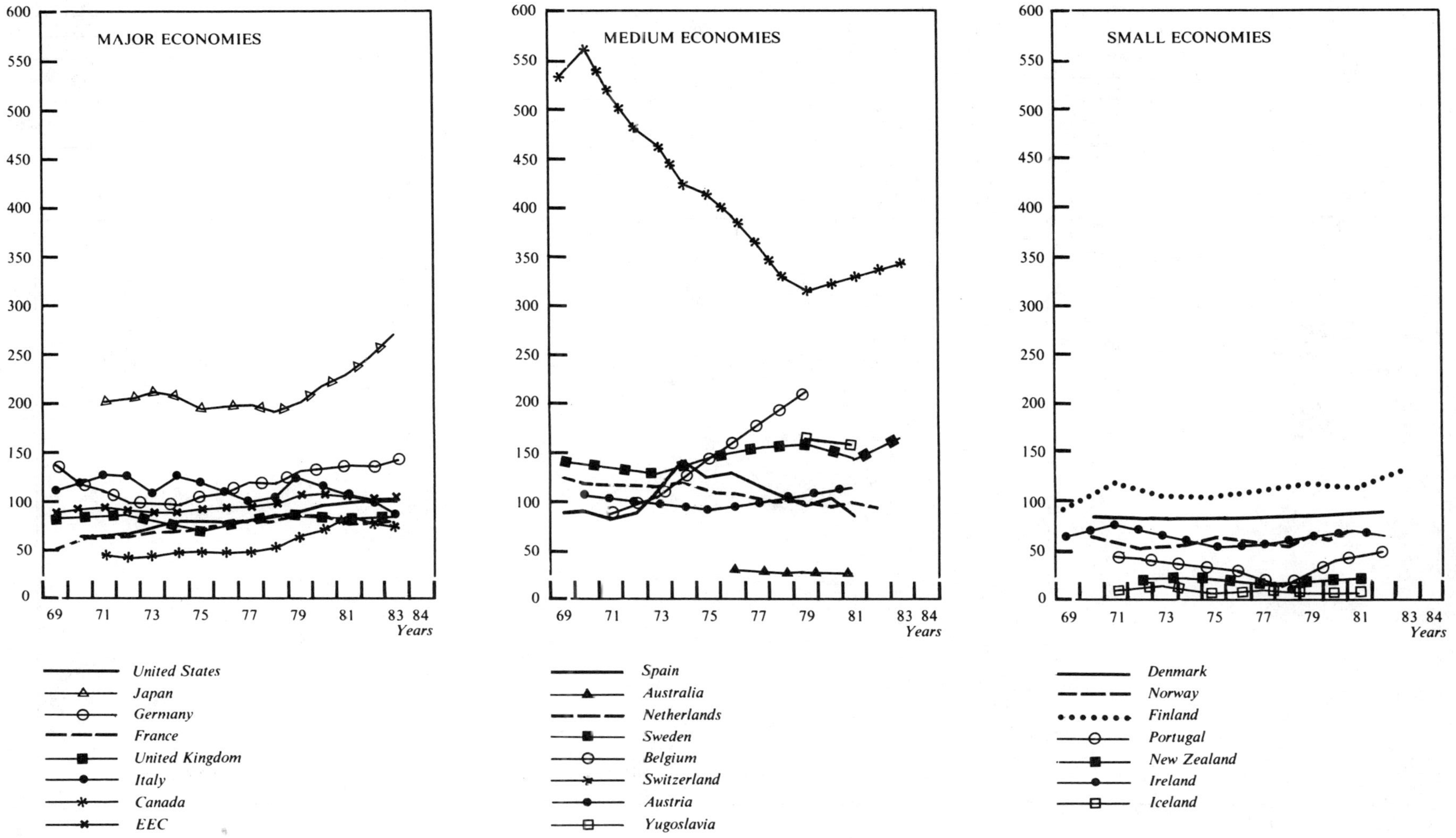

Source: OECD/STIIU Data Bank, November 1985.

Graph 7
Government R&D spending as a percent of total government expenditure*

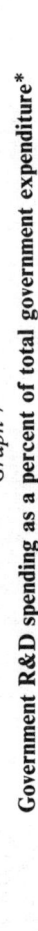

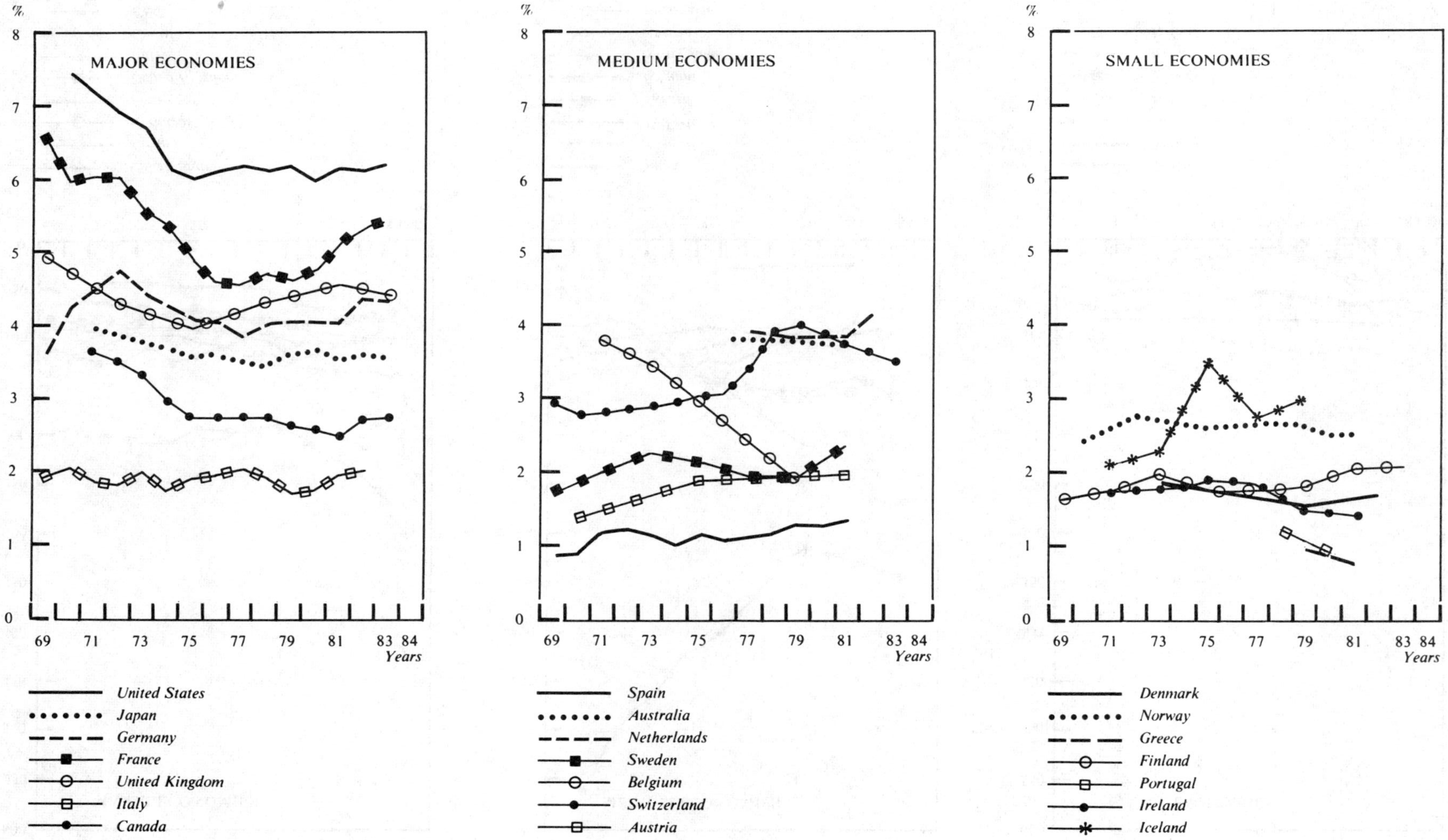

* For details, see note 5 at the end of the report.
Source: OECD/STIIU Data Bank, November 1985.

overestimate the R&D resources in the Higher Education sector – almost two-thirds government financed – the ratio of Japanese public finance in the OECD area would be even more modest. During the early 1980s, the Japanese share of all public-financed R&D in the OECD area will probably level off or even fall after having doubled over the 1970s.

If we take a relative indicator like government-financed R&D as a percentage of total government expenditure[5], a marked difference can be seen in recent years between the large countries (and perhaps Switzerland and the Netherlands) and the other medium and small countries. It would seem that after a long downward trend, followed by a levelling off, R&D in the large countries has generally been surviving budgetary restraint better than the bulk of other government activities. In the other OECD countries, not only was the R&D/total government expenditure ratio lower, but the indicator usually continued either to cruise along at the same rate or decline.

Government's Goals for R&D Spending

Here, we shall change over to a slightly different series of data based on R&D budgets and accounts, which are usually somewhat less comparable than those discussed so far. However, what they lack in accuracy, they make up in political relevance, for these data reflect the main objectives governments had in mind when actually committing money to research and development. Four main aims have been identified: economic development, health and welfare, defence and space, and the advancement of knowledge[6].

For the OECD area as a whole, the predominant objective is *defence and space,* which receives nearly half of all government R&D resources. In fact, the recovery in the volume of government R&D funding in the late 1970s which has been accelerating in the 1980s can be attributed in large part to the shift of United States' policy in favour of more defence R&D.

In second place comes *economic development,* which received roughly one-quarter of OECD government R&D resources. Funds for economic development have levelled off in the early 1980s after a sharp rise in the late 1970s, both reflecting trends in support for energy R&D.

One-fifth of all government funding for R&D was devoted to *advancement of knowledge,* while *health and welfare* accounted for approximately one-tenth of all funds. During the early 1980s, the share of health and welfare R&D in the OECD total is declining after expanding somewhat during the preceding decade.

Taken at face value, these figures give a distorted picture of the "typical" OECD country, because the trend for the OECD area is largely governed by the policy of the United States and one or two other large countries. In order to reduce the distortions due to the effect of size, it is useful to make a cross-country comparison in terms of R&D resources per capita.

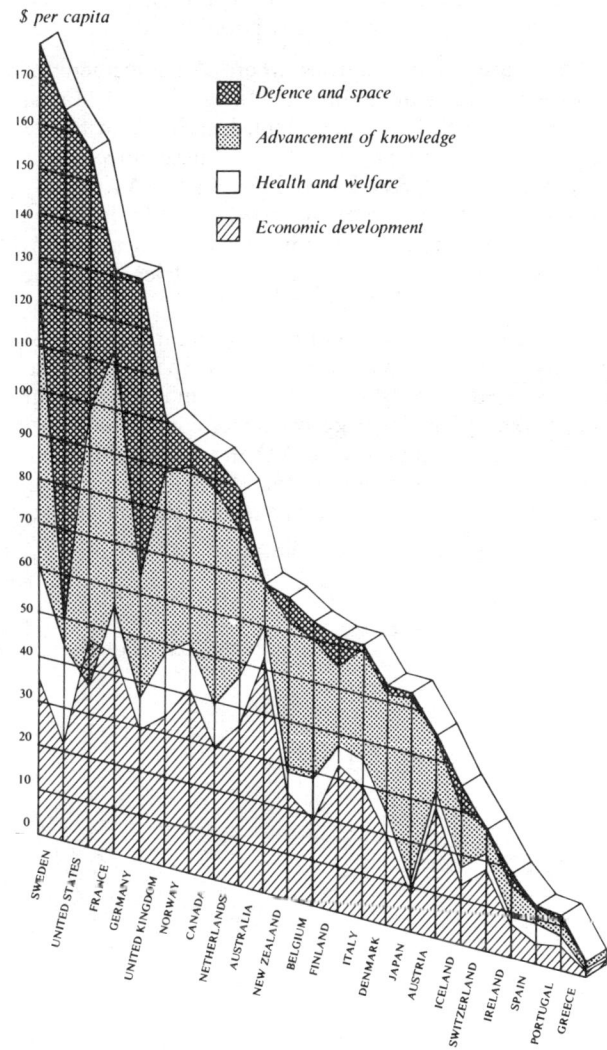

Graph 8
Public R&D funding by socio-economic objectives per capita – 1983
In current dollars using current PPPs

Source: OECD/STIIU, November 1985.

Seen from this angle, the "typical" OECD country had a per capita government expenditure on R&D of about $65 in 1983. Of this, over 40 per cent went to Economic Development, 40 per cent to the Advancement of Knowledge, 10 per cent to Health and Welfare and only less than 10 per cent to Defence and Space (see Annex, Table 5).

When the OECD countries are examined one by one, however, it becomes evident that in most countries Advancement of Knowledge takes first place, coming well ahead of all the other objectives. This is hardly surprising when we realize that, by convention, this class contains all the government support for university research financed from the general budgets from the Ministry of Education. These funds will ultimately be spent on R&D relevant to other aims such as human health or agriculture, although the government cannot actually direct them to such projects when planning its R&D. These grants, known as public GUF, are excluded for Switzerland and the United States where they are provided by provincial government.

There are some countries where Advancement of Knowledge is not the main aim. In the United States, France and the United Kingdom, Defence and Space take the lead. Per capita spending on these objectives is also significant in Sweden and Germany. As already mentioned, Defence and Aerospace R&D lead the growth in the United States but support for these aims is also expected to be up in most countries with the sharpest upturn in Defence in Sweden.

In one or two very small OECD countries, highest priority is given to economic R&D and especially to that part oriented towards agriculture and industry. At the end of the 1970s, most governments committed more money to infrastructure R&D than to agriculture but the situation has changed in the 1980s as government funds for energy R&D have declined in the majority of countries. The main exceptions are France, Canada and Australia. About half of OECD governments have been increasing support to agriculture and industry R&D in recent years.

Per capita spending on R&D earmarked for Health and Welfare is generally fairly low, reaching its highest levels in Sweden, the United States and Norway. Only about half of Member governments increased their specific support for Health and Welfare R&D in the early 1980s.

Who Receives Government R&D Financing?

In 1981 approximately one-third of all public-financed R&D in the OECD area was actually carried out in government laboratories, one-third by industry and one-third by the universities. During the early 1980s the share going to industry has generally been increasing at the expense of higher education, reflecting changes in the objectives of government R&D funding outlined above and varying rates of growth in the receiving sectors.

Graph 9
Public R&D expenditure by sector of performance, 1981

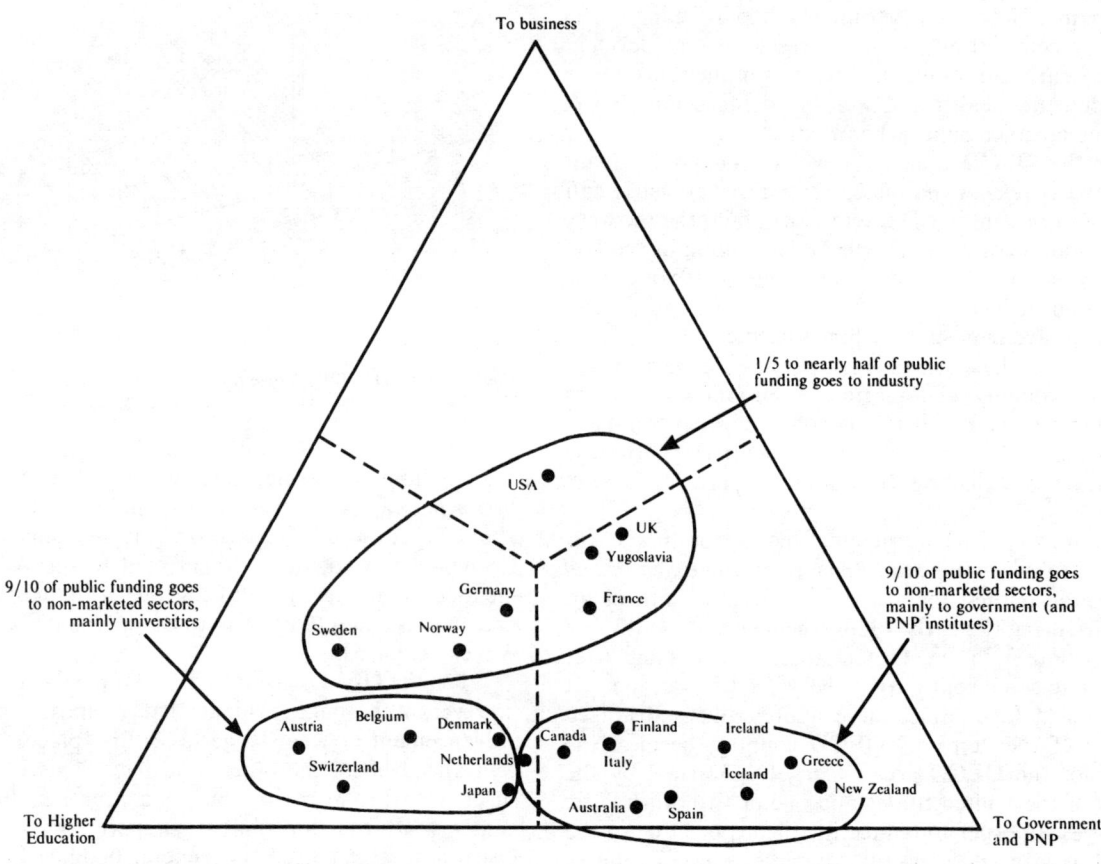

Source: OECD/STIIU, November 1985.

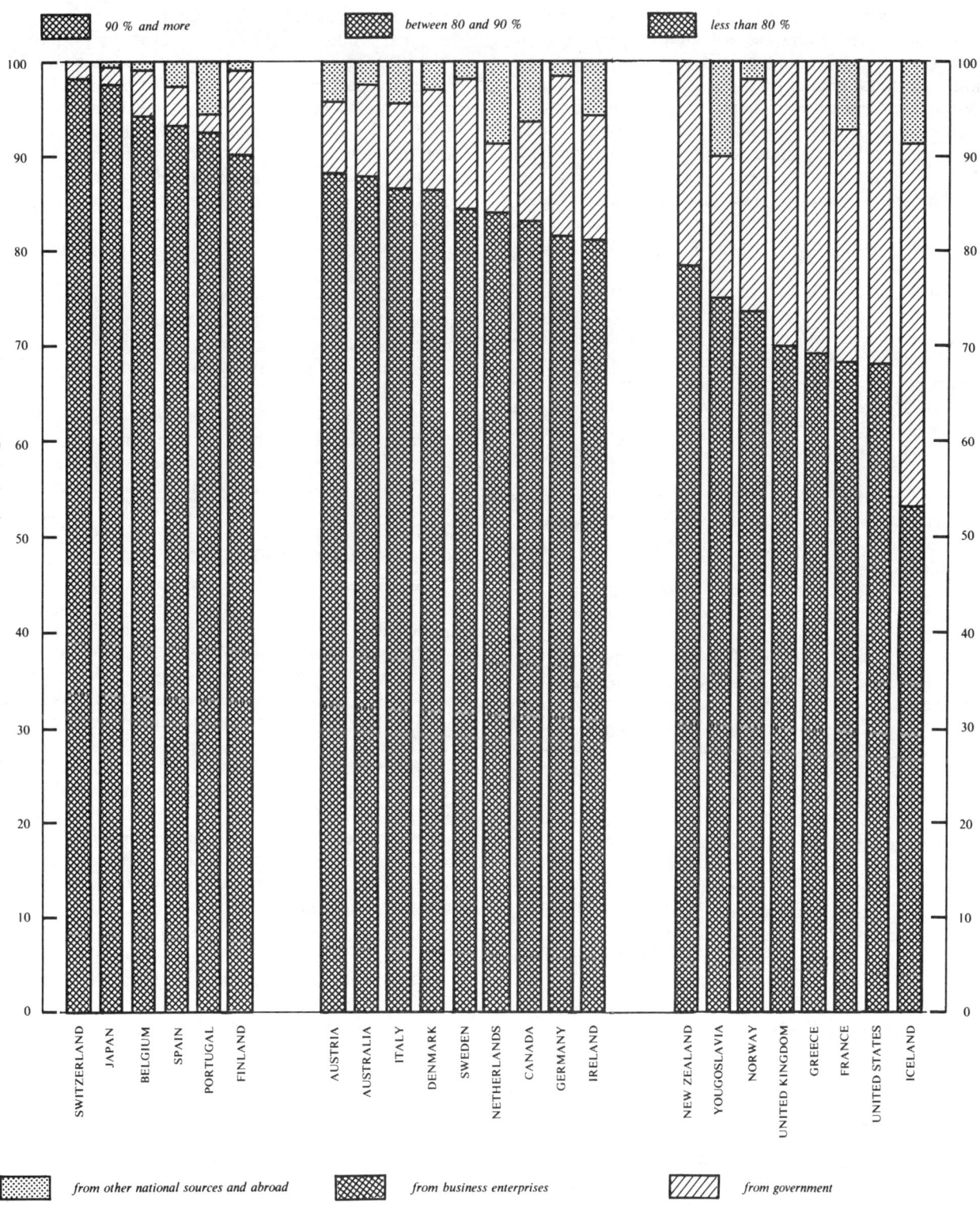

Graph 10
R&D expenditure performed in business enterprise sector by source of funds, 1981
As percent of total expenditure

Source: OECD/STIIU, November 1985.

This overall pattern is, in fact, a composite of quite different types of trends. For the sake of simplicity, they can be grouped into four categories (see Graphs 9 and 10): the United States pattern, the Japanese pattern, the standard and the industrialising patterns[7].

In the *United States pattern,* industry receives the largest share of government funds earmarked for R&D and is given far greater precedence than in the standard or the Japanese pattern, despite the fact that business itself spends heavily on R&D. Even in the less favourable years, industry's share never fell below two-fifths of Federal R&D expenditure. In 1981, universities and government laboratories each received one-quarter of Federal R&D funds although, for a number of years previously, government laboratories had had a slight edge on universities.

Industrial R&D financed by the United States' government is to be found primarily in such high-tech industries as aerospace, electronics and computers. Without the aerospace and defence programmes, government would finance less than 10 per cent of all industrial R&D. Though funding in these sectors declined during the 1970s, the recovery in defence spending is now resulting in substantial new R&D contracts for industry.

The United Kingdom is similar to the United States whilst France is a less extreme version of this pattern. Sweden and Germany are also in this group though defence and aerospace are less important, and in Germany their role is declining. These two governments put more emphasis on research and development contracts for energy programmes.

In the *Japanese pattern,* industry spends heavily on R&D out of its own pocket and is virtually left to its own devices, receiving less than 5 per cent of government R&D funds. In other words, over 90 per cent of government-financed R&D work is carried out either in university laboratories or else in central and local government research establishments. Japanese Government funds are distributed almost equally between universities and government laboratories.

The same pattern of high business-financed R&D, combined with low government support, can be found in Switzerland and, to a lesser degree, in Belgium and the Netherlands. While industrial R&D has remained buoyant in Belgium, business funding in Switzerland and the Netherlands slackened during the 1970s. The Netherlands government reacted by increasing its support of industrial R&D.

The *standard OECD pattern,* which also applies to the EEC as an area, is roughly midway between the other two: the government puts much greater emphasis on university and government research establishments than in the American pattern but allocates far more R&D resources to industry than in Japan. Transfers to business represent about one-tenth of the total; of the remainder, universities usually get a lower share than government establishments (plus private non-profit institutes).

Five OECD countries – Canada, Italy, Austria, Denmark and Finland – typify the standard pattern of industrial research and development, whether in terms of industry's own spending on R&D (just under 1 per cent of value added) or of government support (10-15 per cent of all governmental R&D financing and the same percentage of total R&D performed in the Business Enterprise sector: BERD). Norway and Yugoslavia are similar though industry receives somewhat higher government R&D support. Little or no funds are channelled to industries working in defence or aerospace.

In Canada, government financing of industrial R&D declined during the 1970s, while industry's own funding finally took off. In Italy, the government has recently taken major R&D initiatives and may even be moving towards the United States pattern.

In the pattern for *very small and industrialising countries,* the lion's share of government-financed research and development is carried out in its own establishments. Little or no payments go to industry for R&D, and universities receive relatively low support. Thus government institutions play an exceptionally large role in these countries' R&D effort.

In New Zealand, Ireland and Iceland (and also in Australia) industry itself spends relatively little on research and development activities, largely because most firms are still in natural resource based industries such as mining, agriculture, fisheries or food, which traditionally depend on government for R&D. Governments are therefore giving high priority to stepping up relevant R&D. Most of this support goes to the natural resource based industries, though they also have schemes to encourage all types of industrial firms to build up R&D programmes. They are also among those countries where the highest proportion of R&D for economic objectives is devoted to agriculture, forestry and fisheries. Ireland seems to be moving away from this pattern towards the OECD norm.

In industrialising countries such as Greece, Portugal, Turkey and Spain, industry spends very little on R&D compared with its productive capacity, though the situation is changing in Spain. Governments devote quite a high share of their total R&D budgets to developing industry and infrastructure, but the sums are very small in relation to industrial value added. Moreover, most of the relevant research is done in government laboratories, and the sums actually paid to industry generally finance only a very small percentage of total Business Enterprise R&D (BERD).

Chapter III

BUSINESS: THE MAJOR R&D SECTOR

Some analysts and economists when discussing science and technology activities look upon the Business Enterprise sector as a performer of research and development rather than as a source of funds. This attitude is probably influenced by the national accounts approach and the key role played by industries, and notably corporate and quasi-corporate enterprises in the creation of national wealth. In the OECD area, industry generates four-fifths of gross domestic product.

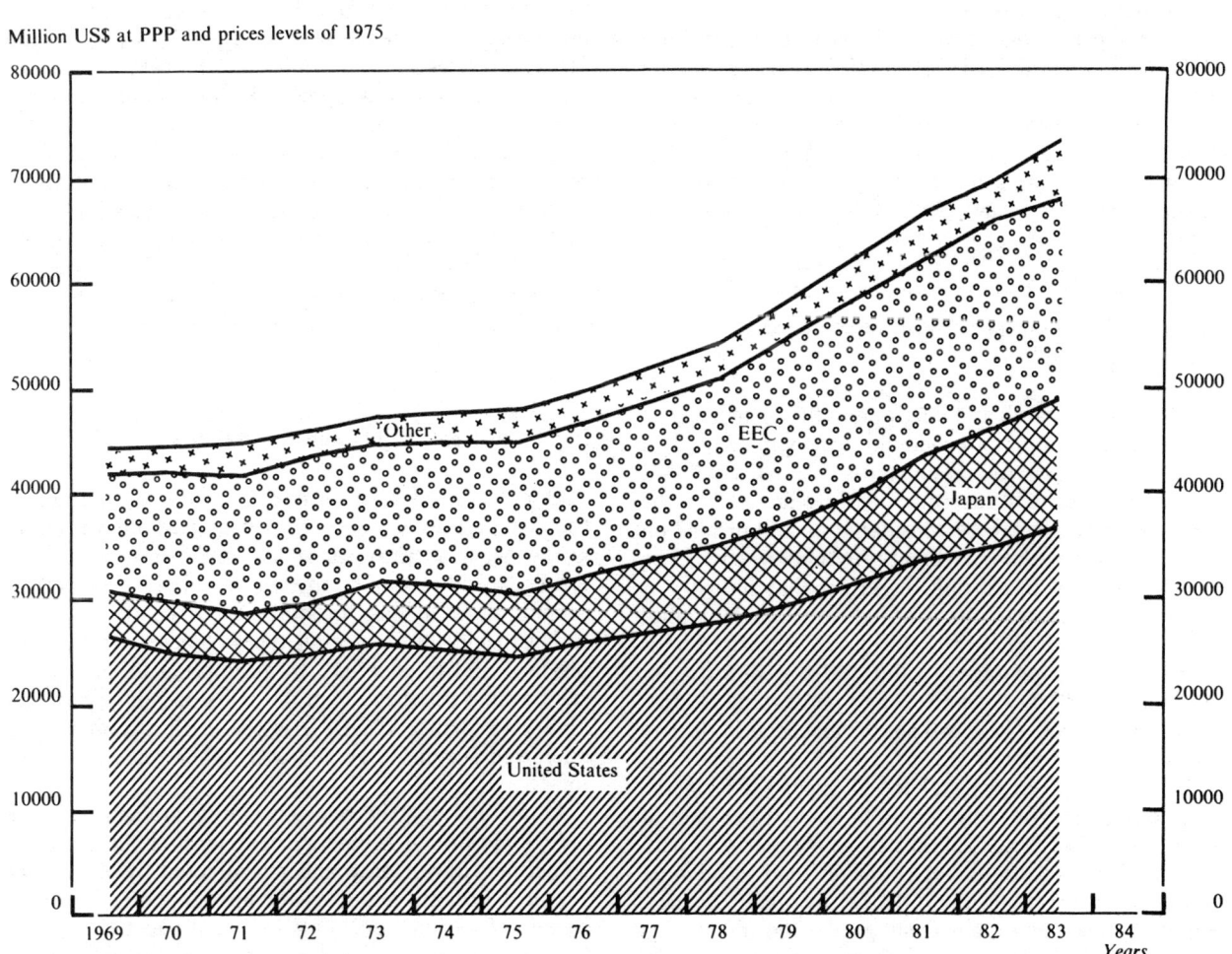

Graph 11
R&D expenditure in the OECD business enterprise sector by area

Source: OECD/STIIU, November 1985.

Table 1.4
Percentages by main OECD areas[1]
When total OECD = 100

	1969	1975	1981	1983
Other	5.3	6.6	6.4	6.2
EEC	26.0	29.9	28.4	27.1
Japan	8.6	12.4	15.1	16.9
United States	60.1	51.1	50.1	49.8

1. For more details see Appendix, Table 7.
Source: OECD/STIIU Data Bank, November 1985.

In 1983, the Business Enterprise sector spent an estimated $126 billion on research and development, which represented two-thirds of all R&D expenditures (GERD) in the OECD area. The geographic breakdown of business R&D is very similar to that for total GERD, except that the United States, Germany and Switzerland do slightly better for industry. In consequence, business R&D is even more geographically concentrated than total GERD.

After a period of recovery which accelerated from 1978/79, industrial R&D in the OECD area is slowing down though still growing. Trends in country shares reflect those already identified in Chapter I with that of Japan continuing to rise whereas the United States fell below the 50 per cent line in 1983. The EEC share also fell to under 28 per cent (from 30 per cent in 1979).

Industrial R&D compared with value added
(See Annex, Table 8)

Although research and development is recognised as an important factor in economic development and in increased productivity, its impact is extremely difficult to measure. As it is generally not "productive" in the short term, R&D can be considered as a "burden" on industry rather than a direct production cost. The simplest way of measuring the effort that industry is making is to resort to intensity indicators such as total expenditure on R&D in the Business Enterprise sector (BERD) as a percentage of value added in industry, or self-financed R&D expenditure as a percentage of gross operating surplus. These will now be looked at in the two sections which follow.

In the OECD area, total R&D expenditure in the Business Enterprise sector came to about 1 per cent of value added in 1983. As in the case of comparisons with total national wealth, larger countries tend to be more R&D intensive than the OECD average. Business Enterprise spending as a percentage of value added is, on average, twice as high in the larger countries as in the smaller ones.

By and large, industrial R&D intensities have grown steadily over the last ten years (notably in Japan, Spain, Sweden, Belgium, Austria, Denmark and Norway).

However, it is necessary to examine trends in industrial R&D spending in the light of structural changes in OECD economies. As is well known, compared with primary activities (agriculture) and tertiary activities (services), the share of the so-called secondary activities (manufacturing, mining, construction and the like) in national economies is decreasing. Whereas in 1960, secondary activities accounted for 40 per cent of gross domestic product in the OECD area, the figure fell to 38 per cent in 1970 and to only 34.8 per cent in 1982. The drop was even more striking in the European Community.

In consequence, if one looks not at the whole of the Business sector but only at manufacturing industries, where nine-tenths of all industrial R&D is carried out in the OECD countries, the intensity indicators of industrial R&D expenditure/value added is two to three times higher depending upon the country (see Annex, Table 8). Thus, in the United States, R&D corresponds to nearly 8 per cent of manufacturing value added and the intensities in the United Kingdom (7 per cent) and Sweden (6 per cent) are also way ahead of the other large countries. The figure is not quite 5 per cent for the other large countries, while the highest figure for manufacturing R&D intensity in the smaller countries is around 3 per cent.

These variations in the global R&D intensity of manufacturing industry reflect a number of factors in each country, notably the balance between high-tech and traditional industries (which is currently being examined in Part II of this report). These intensities are only broad orders of magnitudes and more detailed indicators are needed to explain them.

Industrial R&D Effort and Profitability

One of the more meaningful economic parameters that can be used to assess relative industrial R&D effort is the gross operating surplus GOS, since it enables R&D expenditure to be related to an industry's overall profitability. On this "profitability" depends an industry's capacity to finance all types of investments, both physical and intellectual. In attempting to clarify the relationship between the profitability of industry and its ability to finance R&D, it is useful to examine gross operating surplus together with other related indicators.

Unfortunately, the relevant figures are available for only some fifteen Member countries and the data are sometimes out of date and are affected by significant revisions brought about by the introduction of the new system of national accounts in recent years.

Gross operating surplus as a percentage of value added at factor cost represents that part of production

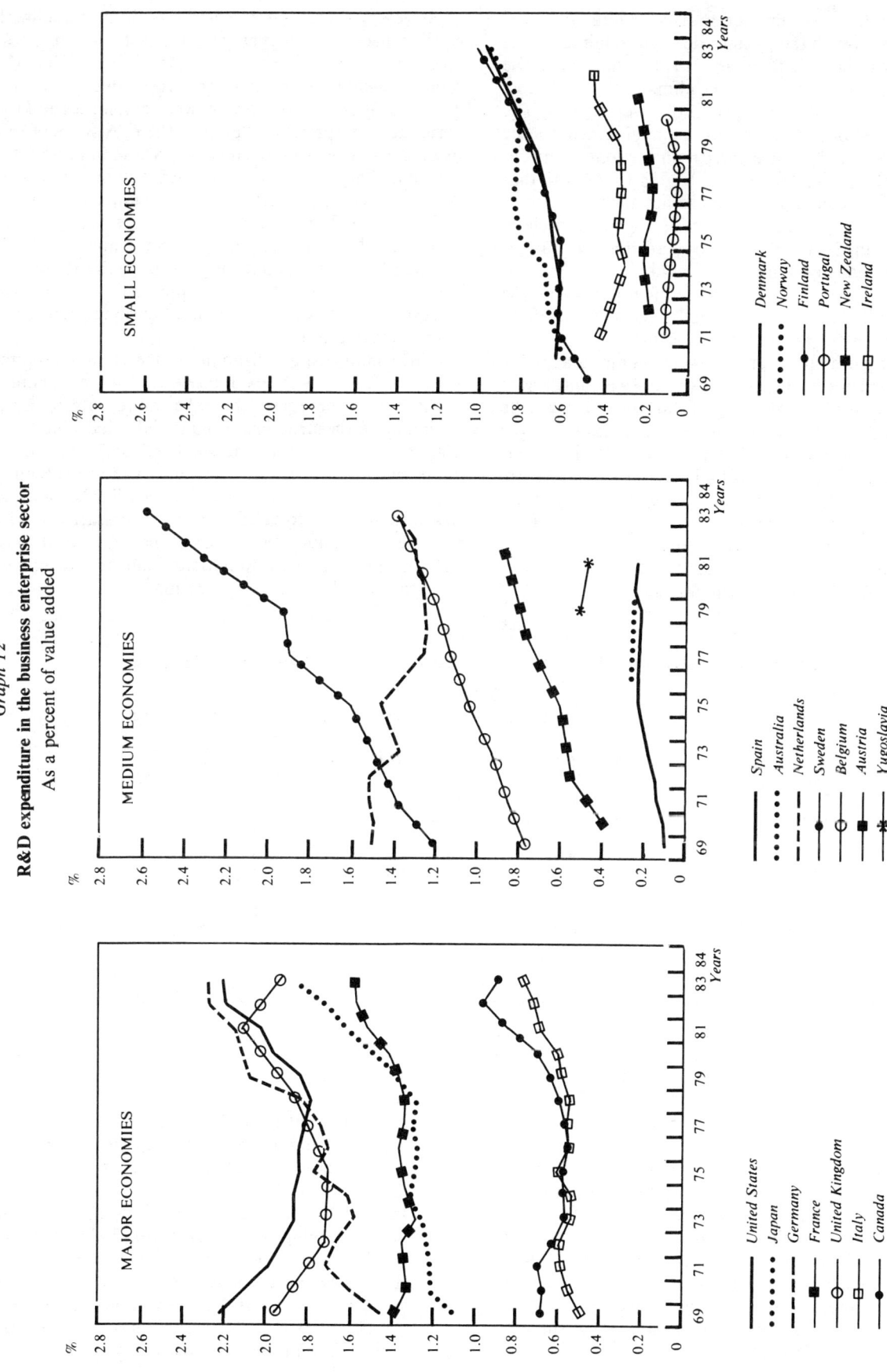

Graph 12
R&D expenditure in the business enterprise sector
As a percent of value added

Source: OECD/STIIU, November 1985.

that is available to remunerate fixed capital (internal and external) with the balance devoted to labour costs. As there can be wide fluctuations from one year to the next depending on the economic climate, this ratio, too, has to be analysed over a period of several years. The share of gross operating surplus in manufacturing value added is on the decline especially in the long term, for example comparing the average for the 1960s with that for 1980-83. A similar decline has not always occurred for the Business Enterprise sector as a whole especially in these countries where the petroleum and natural gas extraction industries engaged a rapid increase in earnings after 1973. Thus, in the United Kingdom, the Netherlands, Norway and Canada, the share of GOS actually grew for the sector as a whole.

On average, a little over 70 per cent of manufacturing value added went to labour costs in 1981 and under 30 per cent to gross operating surplus. By and large, the pattern is the same for the large countries as for the smaller ones for which data are available. However, Japan and to a lesser extent Finland and Portugal seem to be favourably placed with a better equilibrium between the remuneration of capital and of labour. Thus in Japan labour costs represent under 60 per cent of manufacturing value added as against two-thirds to three-quarters in the other countries.

However, these figures should be interpreted with caution because a higher percentage figure for gross operating surplus does not necessarily mean that the profit element is higher, since the gross value added thus generated is not only shared between employees and the corporate enterprises that employ them. A share of this income also goes to pay creditors, whose interest rates are much higher in some countries than in others. Neither should one overlook unincorporated enterprises which are much more numerous in some countries and where the dividing line between profits and salaries is less clearly defined than in large firms. In the absence of such detailed information, these financial ratios provide at best some indication of the profit-earning capacity of manufacturing industry.

When comparing self-financed R&D in manufacturing industry with its relative ability to generate profits, it is necessary to take more international variations in the structure of industrial value added into consideration. The differences in the GOS/value added ratios noted above already go some way to explaining international variations in self-financed R&D/GOS. A low rate of R&D to GOS does not necessarily signify weakness but may, in certain cases indicate the "disposable income" available to national industry within the constraints in home and foreign markets.

Table 1.5

R&D expenditure self-financed in manufacturing industries related to gross operating surplus

In percentage

	1969	1971	1973	1975	1977	1979	1981	1982	1983
United States	16.0	16.8	17.0	17.8	16.0	19.1	24.4	30.2	–
Japan	5.1	5.8	6.3	9.0	9.0	9.3	11.0	–	13.4
Germany	8.0	10.0	9.5	11.8	11.8	14.0	17.8	–	–
France	–	7.7	7.5	9.5	9.6	11.5	–	–	–
United Kingdom	9.4	–	9.5	15.5	–	11.7	22.1	–	–
Italy	5.1	4.0	3.0	3.7	3.9	3.6	4.3	4.3	–
Canada	5.3	5.4	3.9	4.4	4.7	4.7	6.4	10.7	9.2
Australia	–	–	–	–	2.8	3.0	2.8	–	–
Sweden	8.7	13.2	13.1	12.9	31.4	22.6	26.9	–	–
Belgium	–	6.7	7.2	–	–	–	20.8	–	–
Denmark	–	8.1	8.3	8.0	8.8	9.4	8.3	–	–
Norway	5.7	6.5	–	6.7	9.2	6.9	9.3	–	–
Finland	3.1	4.7	4.1	5.0	6.1	5.0	6.3	–	7.0

Source: OECD/STIIU Data Bank, November 1985.

Table 1.5 suggests the following comments:

- In the case of the five large countries plus Sweden, it is in Japan that self-financed R&D weighs the least heavily on gross operating surplus even though such self-financed R&D is the highest of the OECD area (along with Switzerland). However, the ratio is at its highest in the United States and Sweden (remembering that these are also the two countries where GOS is the lowest share of manufacturing value added).

- In other countries self-financed R&D weighs less heavily, which is only to be expected given their smaller industrial R&D efforts and greater specialisation.
- In virtually all the countries, the weight of R&D on the earning of manufacturing industry grew, notably in the United States and Sweden where the R&D corresponded to a quarter of gross operating surplus. In Japan the ratio doubled over the period though "disposable income" was still higher than in the other countries.

R&D by Industry Groups

While the intensity of national R&D efforts in industry as a whole is necessary to obtain a picture of this component of all research and development activities, it is also important to examine trends in the different industry groups within this sector. R&D can be much more vital to some groups than to others, particulary to high technology industries which contribute much more than others to total industrial R&D.

Unfortunately, the amount of industrial detail available in several large countries is declining for various reasons, notably disclosure problems caused by increased industrial concentration. This makes it difficult to calculate OECD totals or examine trends for the different industry groups for the latest years. The following analysis is based on data for 1981. In 1983 results by industry were available for only two-thirds of OECD countries. Table 9 in the Annex shows the share of each industry group in each country's total industrial R&D. Table 10 in the Annex shows each country's share of the OECD total for each industry group.

In general, the relative position of the industry groups changes only slowly as can be seen from Graph 13 which shows the situation in 1975 and 1981 shortly after the two petrol shocks.

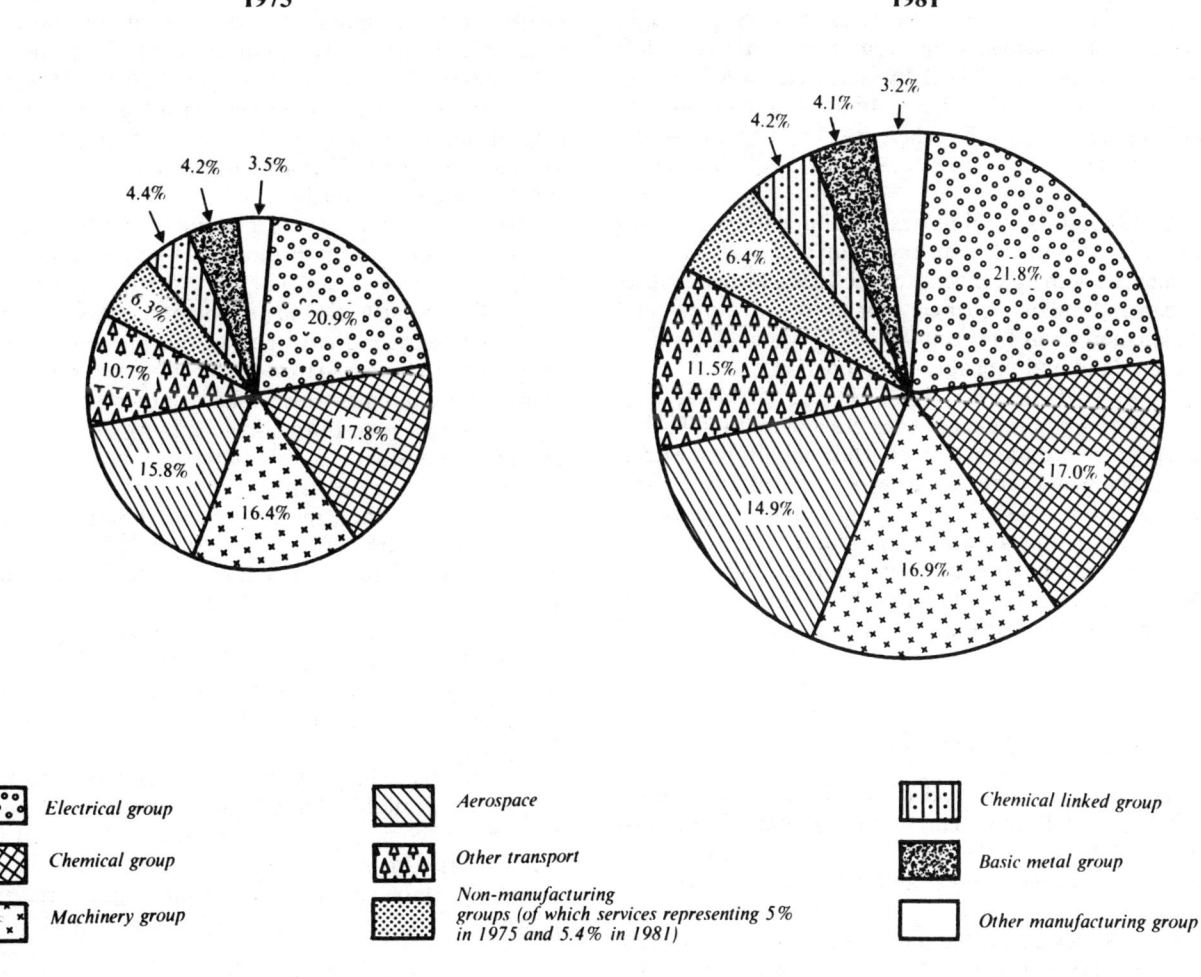

Graph 13
R&D expenditure by industry groups
As percentages of the business enterprise sector

Source: OECD/STIIU Data Bank, November 1985.

The Electrical/Electronics Group (Annex, Table 11)

This group comprises the manufacture of electrical and electronics machinery, apparatus, appliances and components, but does not include data processing equipment (computers) which, according to the International Standard Industrial Classification (ISIC) used by OECD, is included in machinery with office machinery.

This group accounted for more than one-fifth of Business Enterprise sector R&D spending in the OECD area with about $23 billion in 1981, and for one-quarter of the researchers. It financed more than seven-tenths of its R&D out of its own funds, the remainder coming largely from government.

During the past decade, research and development in this group grew at a slightly slower rate than total industrial R&D in the OECD area. It was only after 1975, in fact, that positive growth was registered. Before that, growth had been reduced virtually to zero by the steady decline in the electrical/electronics group in the United States, which hit its lowest level in 1975.

The difference in growth rates, however, brought about a major change in the geographic pattern of R&D in these industries. The EEC countries were able to maintain their 1975 share, that of Japan enjoyed uninterrupted growth whereas the United States dropped back more than for almost any other industry.

By 1981, the United States accounted for over 46 per cent of R&D for this industry group in the OECD area, compared with 31 per cent for the EEC, 17 per cent for Japan and roughly 6 per cent for the other countries (of which more than 1 per cent apiece for Canada, Sweden, Switzerland and probably Belgium).

From 1979 to 1981, R&D in the electrical/electronics group in most countries grew more rapidly than during the four previous years, while only Italy, Switzerland, Spain and above all Norway recorded a drop. Data available for 1983 for a significant number of countries confirm that this growth is continuing.

Except in one or two small countries, this upturn in R&D was essentially financed by business funds with government support following the trend.

Chemicals (Annex, Table 12)

This group is in second place with some $18 billion spent on R&D representing 17 per cent of business-financed R&D in the OECD area. It comprises the chemical, pharmaceutical and petroleum industries.

In 1981, the area total breaks down as follows: the United States accounted for 41 per cent of total R&D, the EEC for 33 per cent, Japan for 16 per cent and the other countries for about 10 per cent. The differences of R&D effort among the major spenders are less pronounced than in the electrical/electronics group. In recent years, Japan has been gaining ground at the expense of the EEC with the shares of the United States and other countries remaining stable.

In the chemicals group, the firms themselves finance the major part of their research and development (more than nine-tenths). The pattern of funding has remained astonishingly stable compared with the electrical/electronics group, which depends more heavily on government financing, especially in countries with major defence and space programmes.

The growth curve in most countries is also more regular than in the electrical/electronics group as trends in the individual industries tend to cancel each other out. Generally speaking, however, the petroleum and pharmaceutical industries have increased their R&D expenditure faster since 1979 than has the chemical industry per se.

Machinery Group (Annex, Table 13)

This group, which includes the non-electrical machinery, office equipment (including computers), precision instruments and equipment industries, shares the same rank as the chemicals group: in 1981, it spent over $17 billion on research and development. This was equivalent to nearly 17 per cent of all business R&D expenditure in the OECD area, a share that has been maintained since 1975 thanks to business sources which supply more than four-fifths of all R&D spending.

The United States takes a commanding lead in this industry group, putting it much farther ahead of other OECD countries than in the electrical/electronics group. This is due in part to its computer industry, which alone accounts for almost half of this group's expenditures in the United States. However, precision instruments and equipment also take a larger share in the United States (one-third of the group) than in other countries (one-tenth in the major European countries and one-quarter in Japan).

In Japan, too, these industries have shown strong and steady growth: its share of total business R&D expenditure jumped from 3 per cent in 1969 to 12 per cent in 1981.

In Germany, this group had a period of rapid and sustained growth during the 1970s, doubling its share to reach 16 per cent of national industrial R&D, followed by an abrupt slowdown in 1981, only maintaining the 1979 level.

As for the other countries for which comparable figures are available, there was a marked turndown in the French computer industry's share of the OECD R&D total. On the other hand, Norway, Canada and especially Italy showed substantial though irregular growth.

Aerospace (Annex, Table 14)

Aerospace, in the 1960s the most important R&D performing industry in the OECD area, is now in fourth place. It is the most geographically concentrated of all the industry groups. The United States is synonymous with aerospace R&D, accounting for 11 of the $15.5 billion spent on such R&D in the OECD area in 1981 (15 per cent of total industrial R&D). Next come three other countries with major aerospace programmes: France, the United Kingdom and Germany, each spending at most one-tenth of the United States' outlay. Italy (whose spending is increasing rapidly), Canada and Sweden are the other members of the "Aerospace Club". Japan's share, according to official figures, is negligible.

In contrast to, say, the chemicals group, which draws mainly on its own funds, research and development in the aerospace industry relies heavily on government funding, notably through procurement contracts and subsidies. In the four countries with major programmes, government pays for about 70 per cent of the R&D.

Since 1981, this industry's share has risen slightly in the United States and Canada, while in Italy there has been substantial growth. The percentage financed by government grew significantly in Canada and Italy and was also up in France and Germany.

Other Transport Equipment (Annex, Table 15)

This group of industries includes motor manufacturing, shipbuilding and the manufacture of such other transport equipment as motorcycles, cycles, rolling stock, etc.

In the OECD area, this group spent a total of almost $12 billion on R&D in 1981, which represented more than 11 per cent of all business expenditure for R&D. The rate of growth in R&D spending fell back sharply from 1979 onwards after having been one of the highest during the 1970s. Nearly all research and development activities in the group are self-financed. The respective contribution of the various industries within the group are: motor vehicles, 90 per cent, and 5 per cent each for shipbuilding and other transport materials.

The biggest change in this group has been the rise of Japan which by 1981 was responsible for over 20 per cent of the OECD total, largely thanks to its motor vehicle industry. Between 1979 and 1981, Japanese R&D in this group rose from the equivalent of 70 per cent to 90 per cent of EEC spending.

Chemical-linked Industries (Annex, Table 16)

This group encompasses three families of industries: food, drink and tobacco; textiles, footwear and leather; and rubber and plastic products. Together, they invested more than $4 billion in R&D in 1981 or some 4 per cent of the OECD total. This puts the group on a par with the basic metals industries. The food industries represent 45 per cent of all expenditure of the group, followed by the rubber and plastics industries (40 per cent) and much further behind, the textile and allied industries (15 per cent). Nine-tenths of this money came from private sources.

This group usually fell back as a percentage of all industrial R&D, particularly in Italy, Sweden and Ireland. This was also the group where the United States' share of the OECD total declined the most obviously.

Germany and Japan were the only two countries where R&D was relatively buoyant, compared with the OECD area as a whole, both in the group and in the constituent industries. Growth was particularly marked in Japan overall and in the textile industry. By 1981 the Japanese textile industry was spending twice as much on R&D as its United States counterpart whereas they had been level pegging in 1975.

Also worth noting is the substantial effort made in New Zealand and Ireland, where this group of industries perform about one-quarter of all industrial R&D, six times the OECD average for the group. In both, the biggest R&D effort is being made by the food and drink industries.

Basic Metals Group (Annex, Table 17)

As a rule, this group of industries, which includes ferrous, non-ferrous metals and fabricated metal products, is not strongly committed to intensive R&D, for it can take up to a generation to develop a new technology. In 1981, the group spent $4.3 billion in the OECD area, representing roughly 4 per cent of total business R&D expenditure.

Annual growth rates for individual industries have been very uneven, and the group has maintained its share only because these erratic movements have in most countries offset one another. In all these industries, nearly nine-tenths of R&D funds come from business, down slightly since the 1970s. In the OECD area, ferrous metals account for two-fifths of the group's expenditure on R&D, while fabricated metal products account for one-third and the non-ferrous metals industry about one-fourth.

Japan is responsible for about 40 per cent of OECD R&D in the ferrous metals industry, spending as much as the United States and Germany combined. Sweden and Switzerland are also significant for ferrous metals R&D. The EEC plays a more important role in fabricated metal products largely because of the R&D efforts of Germany (which spends almost as much as Japan) and, at a lower level, of Belgium and Ireland. As might be expected, R&D on non-ferrous metals is particularly significant in Canada and Norway and also in Italy and Finland.

The Service Industries (Annex, Table 18)

In 1981 this group devoted some $5.9 billion to R&D, or 5.3 per cent of total industrial R&D expenditure in the OECD area. This was after a steady rise during the 1970s, R&D investment in this group growing faster on average than that of industry as a whole.

In almost every country, this increase has been primarily due to business finance, exceptions being the United States, Germany, Sweden, Norway and Denmark, where government finances at least one-third of the total.

The service industries rank way above average in most of the medium and small countries, whereas they are much less well placed in the large countries and in Switzerland, Finland and Iceland.

Though the group is composed of six distinct industries, there is so much variation in the way countries report data to the OECD survey that it is not possible to make reliable international comparisons.

Chapter IV

UNIVERSITY RESEARCH AND BASIC RESEARCH

In the OECD area, university research accounts for only about one-sixth of all R&D funds. On the other hand, an average of one out of every four researchers works in university laboratories, though in a good number of countries, small or large, the proportion can be significantly higher-reaching or even exceeding half the nation's researchers as in Australia, Belgium, Austria, Switzerland, Ireland, Italy and Spain (see Annex, Table 19).

Despite the special problems of evaluating the resources devoted to R&D in the Higher Education sector, which is the weak point in national measurement systems, the three following questions will be answered as far as possible:
- What is the geographical distribution of university research in the OECD area and how has this changed in recent years?
- To what fields of science are the university R&D resources devoted?
- What is the rough geographical distribution of basic (and applied) research between the main OECD zones?

Distribution of university research

In 1983 some 29 billion dollars was spent on R&D in the Higher Education sector in the OECD, employing the full-time equivalent of 375 000 researchers. Both these figures have been calculated after adjustment of the Japanese data to make them more comparable with those for other Member countries (for further details see note 4). These adjusted data give a geographical breakdown of university R&D which is more similar to the pattern for the other sectors examined in previous chapters than the unrevised statistics. This is because the latter include both the teaching and the research activities of Japanese university staff which gives the impression that there are as many researchers in the Higher Education sector in Japan as in the United States, Germany and France combined! This difference in survey practice has been sufficiently confirmed by the competent national authorities to allow an adjustment to be made.

According to these adjusted data, the geographical distribution of research resources between the United States, Japan, the EEC and the "other" countries is much more even for the Higher Education sector than for the others, especially Business Enterprise. This reflects the fact that the relative amount of resources devoted to university R&D (spending as a percentage of GDP) is generally as high in the EEC and the other countries as in the United States or Japan. Only in the very small countries, Italy, Spain and Yugoslavia is the figure well below the OECD norm of 0.35 per cent of GDP. The EEC countries combined are actually the largest employers of university researchers in the OECD area, suggesting that university research (and even university basic research) is a European phenomenon.

Spending on R&D in the Higher Education sector is generally falling back compared with national totals. Such funding was for many years largely a function of expected student enrolments (at least that part associated with general university funding) and this still applies in one or two countries. The reasons for the relative decline in university R&D are many and complex. They include:
  i) the slowing down in the rush of young people into higher education, particularly into long courses,
 ii) the declining share of natural sciences and of engineering in total enrolment and graduation, and
iii) especially at post-graduate (ISCED 7) level.

Comparisons by Field of Science (adjusted data used)

Roughly 82 per cent of university R&D in the OECD area is devoted to the natural sciences and engineering and the remaining 18 per cent to the social sciences and humanities (Graph 14). Here again, the United States and Japan are at opposite extremes: whereas the United States devotes about 95 per cent to the natural sciences and engineering (NSE) and only 5 per cent to the social sciences, the break in Japan is about 60 per cent and 40 per cent respectively. In consequence, Japanese university spending on R&D in the natural sciences and engineering falls well behind that of the EEC countries whereas the United States spends the same as the other two combined.

Looking at each group of scientific discipline separately at all-OECD level, it is clear that the natural sciences and agricultural sciences occupy first and last place respectively, which simply reflects the structure of university research in most individual OECD countries. However, the other major fields – engineering, medical science and the social sciences and humanities – all have a fairly similar share due to the preponderant influence of the United States on figures for the natural sciences and of Japan for the social sciences.

In the *natural sciences,* the United States takes the lead with over half of the total. Japan's share is clearly smaller than that of Germany, France, the United Kingdom and even Canada, despite the fact that it spends much more on university research overall.

In the case of university expenditure on *engineering* and even more so on *medical sciences,* the shares of the United States and the EEC are much closer, with the EEC ahead for the medical sciences. The "other" countries are also well placed in this field. It is almost as if there were a consensus approach to these "traditional" fields where the short and medium social or economic

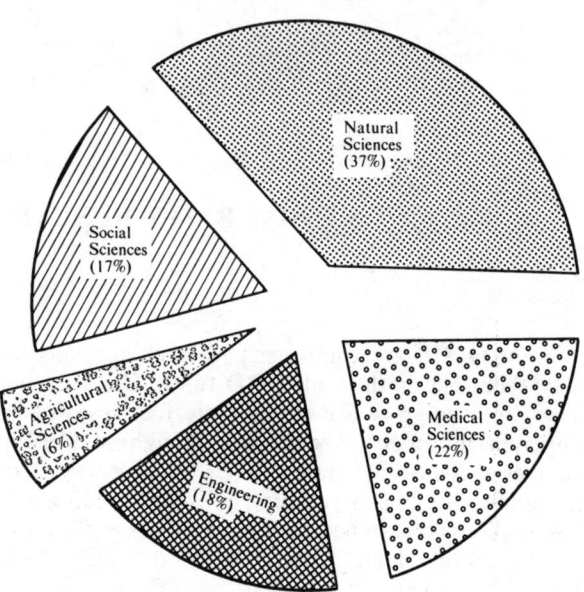

Graph 14
High education R&D expenditure by scientific fields in the OECD area, 1981

Source: OECD/STIIU Data Bank, November 1985.

Graph 15
Higher education R&D expenditure by field of science and main OECD area, 1981
When each field of science = 100

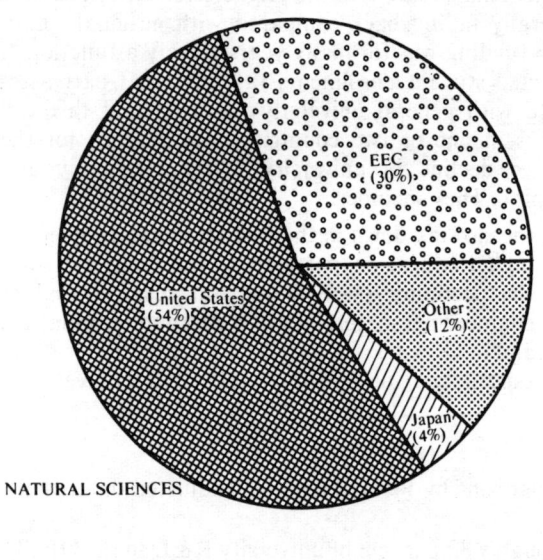

NATURAL SCIENCES

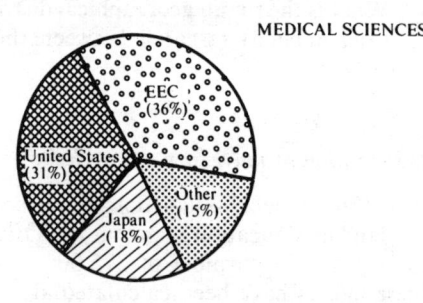

MEDICAL SCIENCES

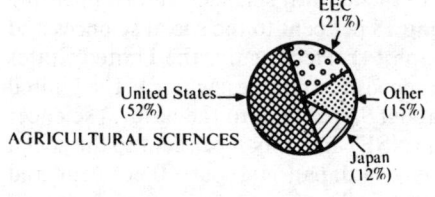

AGRICULTURAL SCIENCES

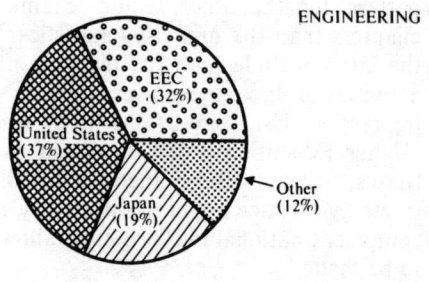

ENGINEERING

Source: OECD/STIIU Data Bank, November 1985.

effects are easier to appreciate than those of the national sciences or of the social sciences and humanities.

Agricultural science generally accounts for only a very small portion – about 5 per cent – of university R&D funds in both small and large countries. The internationally predominant role of the United States is striking and the "other" countries are well placed (notably Sweden, Australia and Yugoslavia), agricultural science in universities having an average intensity nearly double that of the EEC and Japan.

It should be remembered that the above comparisons cover only R&D in the Higher Education sector. They give some idea of the characteristics and relative strengths of each country's universities in R&D in the major fields in the OECD. They do not cover all R&D in each field in the country, nor even in the total "public" sector (universities + government laboratories), due to the differences in institutional organisation in R&D as a whole in the United States, Japan and the EEC.

National Efforts in Basic Research
(adjusted data used)

Because of its potential impact on the economic and social life of a country, the question of basic research periodically gives rise to heated debate in science policy circles. Though this impact is extremely difficult to assess, the first question that is inevitably posed is: what types of basic research should be pursued to reach which aims and, of these, which should be given priority? Another, often unspoken query is: how much basic research should the country be doing compared to other countries? Without entering into a detailed analysis, the following observations, based in part on rough OECD estimates, try to draw some comparisons on basic research between major geographical areas and between the various sectors of performance.

In the OECD area, estimated current expenditures on basic research in the natural sciences and engineering amounted to approximately $21 billion in 1981. Out of every $100 spent on these sciences, roughly $15 went to basic research, over $40 to basic cum applied research and nearly $60 to experimental development. This pattern is more in line with that of the United States and of Japan as, in the EEC as a whole and in "other" countries, the share of research (fundamental and applied) and that of development were more or less equal (50/50).

If we re-examine the $15 for basic research, the United States contributes nearly $7, the EEC $5, Japan and the other OECD countries $3. Of the $42 going to basic and applied research, the United States contribute $19, the EEC $14, Japan $5 and the rest $4. As for experimental development's $58, the United States accounts for $32, the EEC $14, Japan $8 and the

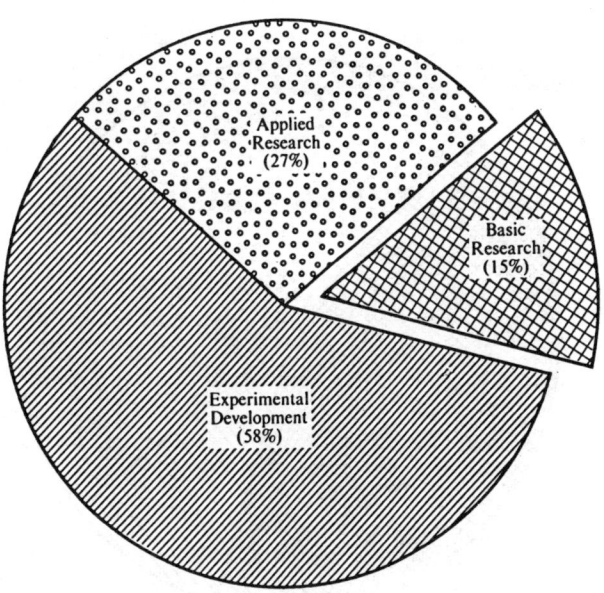

Graph 16
R&D by type of activity in the OECD area, 1981

Source: OECD/STIIU Data Bank, November 1985.

"other" countries $4. The EEC countries seem to be better placed for research whereas the United States is particularly strong in development spending. Japan falls between the two, playing about the same role in research as in development at OECD level.

As might be expected, it is the universities that perform the major share of basic research in the OECD area, accounting for some 60 per cent of funds spent on such research. The balance of basic research is roughly evenly divided between governmental and industrial research laboratories with a minor contribution from PNP institutions.

In the EEC as a whole, university basic research is at a comparable level to that of the Unites States. However, the roles of government laboratories and industry in basic research expenditures vary considerably between geographic zones. The EEC countries combined have a particularly strong basic research effort in government laboratories spending more than the United States and Japan combined; small and medium countries also stand up well to international comparison at the government laboratory level. This is not true for industrial basic research where the balance is clearly in the United States' favour and which is also gaining ground in Japan (see Graph 17).

The preceding observations give some idea of the effort that various countries are making in favour of basic research in government, industry and universities.

Graph 17
Basic research by main OECD areas and sector of performance, 1981

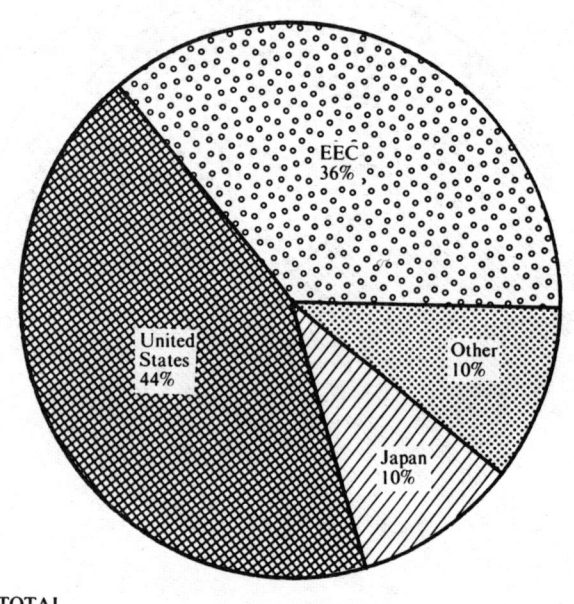

TOTAL

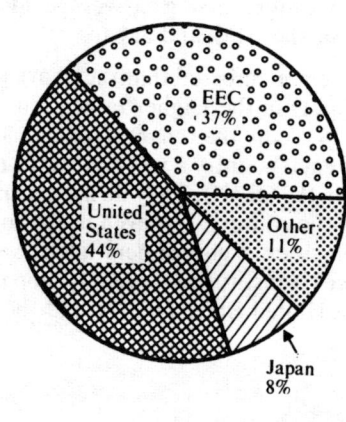

HIGHER EDUCATION

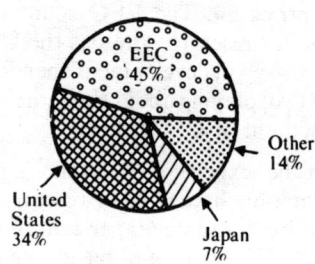

GOVERNMENT

ENTERPRISES

Source: OECD/STITU Data Bank, November 1985.

They can be put in perspective by one last observation. It is often assumed that basic research in a given scientific field can be transferred fairly easily from one economic sector to another. However, a recent study carried out among Japanese industrialists gives food for thought. Of the industrialists interviewed, over 60 per cent stated that they could not use the results of basic and applied research carried out in university laboratories, while only 27 per cent said they found them useful. As for results from government laboratories, nearly 49 per cent of the industrialists stated they were not useful, while 34 per cent said they were. It would seem that the breach between industry and public laboratories and specially universities continues, indeed, to be wide.

Part II

TECHNOLOGICAL PERFORMANCE AND INDUSTRIAL COMPETITIVENESS

INTRODUCTION

This second part of the report sets out to evaluate countries' situations regarding the production and international dissemination of technology and to relate these to their industrial competitiveness.

Chapter I discusses trends in the technological situations of OECD countries in the 1970s and early 1980s on the basis of two types of indicators of the "output" of the S&T system: patents and the technological balance of payments, both of which are now accepted tools for this kind of analysis. It should be noted from the outset that there is no unique measure of technology which is meaningful for all types of economy so several variables are used including some of the R&D data already discussed in Part I.

In Chapter II, the analysis is broadened to take in the relationship between technological performance and industrial competitiveness, the basic idea being that the export performances of high R&D intensity industries are not independent of the factors of competitiveness which may also affect the other industries. Similarly, competitiveness and industrial performances in general are not unrelated to research and development and its effectiveness, so the approach will be broadened to cover all industries in such a way as to establish qualitative rather than quantitative links between R&D and industrial competitiveness.

Part I of the report was able to cover virtually all Member countries as response rates to OECD R&D surveys are very high. The types of data used in this second half vary considerably in availability and quality giving less coherent coverage.

Chapter I

TRENDS IN THE TECHNOLOGICAL SITUATIONS OF OECD COUNTRIES

The technological situations of OECD countries and the relevant trends during the 1970s and early 1980s will be discussed first as "producers" of technology and then as "centres" for the international dissemination (or receipt) of technology.

The analysis will be based on two sets of data, patents and the technological balance of payments which, as was noted in the general introduction to this report, are now broadly accepted as indicators of the "output" of the S&T production system and thus fall between the "input" (R&D) data discussed in Part I and the "impact" indicators (production of and trade in technologically intensive products) to be discussed in the next chapter.

At first sight, these two sets of data correspond to the two phases of production (patents) and diffusion (TBP) of technology.

However, this breakdown is less clear-cut than it seems in that data on international patent applications [whether external or foreign applications or the filing of patents under the European Patent Convention (EPC) or the International Patent Co-operation Treaty (PCT)] can be used to analyse the situations of countries both as producers of technology and as centres of dissemination.

First, external or international applications for patents are not possible without the original production of technology and are, moreover, considered to be better indicators from this point of view than domestic applications. Secondly, external and international applications may provide a defence against competition on external markets and are, therefore, a preliminary to penetrating these markets by means of trade or granting of licences. Changing trends in international and external patent applications are, thus, one set of indicators measuring the roles which the various countries play in the international dissemination of technology.

The section on patents does not, therefore, mark a distinction between the two activities (production and dissemination of technology) but rather indicates their interdependence.

PATENTS AS MEASURES OF THE PRODUCTION OF TECHNOLOGY

Patents are a science and technology output indicator but, owing to their specific characteristics, cannot be regarded as accurate measures of the numbers of inventions. Data on patents can, nevertheless, be used to assess the situations of the various economies as producers and users of technology. Moreover, the existence of international patent systems (EPC, PCT) and of foreign or external applications within national systems means that series are available which can give some indication of how the various economies fare in international dissemination of technology.

Characteristics of the Data

Various criticisms are often levelled against patent data to show that they are not good measures of invention, industrial innovation or of the output of R&D.

First, patent data are not comparable with each other for a number of reasons. They vary in inherent quality. National patent offices make different charges and impose different conditions. The propensity to patent varies between industries depending on the ease of imitation and the strategy of the firms. There may be breaks in series caused by changes in any of the above. Second, the set of patents and of inventions are not coterminous as some inventions are not patentable and others are not patented. Thus, there is no one-to-one relationship between patents and inventions which would allow patents to serve as a standard measure of a nation's inventiveness.

Such criticism, however, simply overlooks the fact that patent systems were never intended to measure inventiveness but have two essential purposes:

- to ensure that the inventor and innovator has the exclusive use of the patented technique during a given period; and
- to give a certain amount of publicity to technological innovations under way.

What matters of course from the applicant's point of view is the assurance of protection and exclusiveness and the filing of a patent can be seen as prerequisite to committing funds to the industrial and commercial launch. Patent applications are located at the interface between the formulation of new technical information and the industrial process; it is not surprising to find that there are correlations on the one hand between R&D expenditure and patents at a sufficiently disaggregated level and, on the other hand, between the trend in patents' statistics and industry's introduction of new products and processes.

The statistics compiled from national data by the World Intellectual Property Organisation (WIPO) and which, from 1978 onwards, are supplemented by EPC and PCT patents, will be analysed in the following in the context of the OECD area.

There are two major types of patent data:

- applications, i.e. patents filed, and
- grants, i.e. patents obtained.

Not all applications result in the grant of a patent. In OECD countries, taken together, the average ratio between grants and applications ranged from 57 per cent to 46 per cent between 1965 and 1982 (WIPO data only, see Table 27 in Annex). This ratio varies from country to country and over time (from 13 per cent in the Netherlands in 1970 to almost 100 per cent in Belgium in 1980). Given these variations, which depend on the existence and severity of the examination process in each country, it is preferable to use the applications series.

Patent series are themselves broken down into four categories: *national* applications/grants (total filed or granted in a country), which consist of *domestic* applications/grants (filed by/granted to residents) and *foreign* applications/grants (filed by/granted to non-residents). Lastly, *external* applications are those filed by the residents of a country outside its frontiers (the sum of foreign applications, thus, equals the sum of external applications for all countries taken together). The diagram below illustrates the relationship between these four categories for a given country, named A:

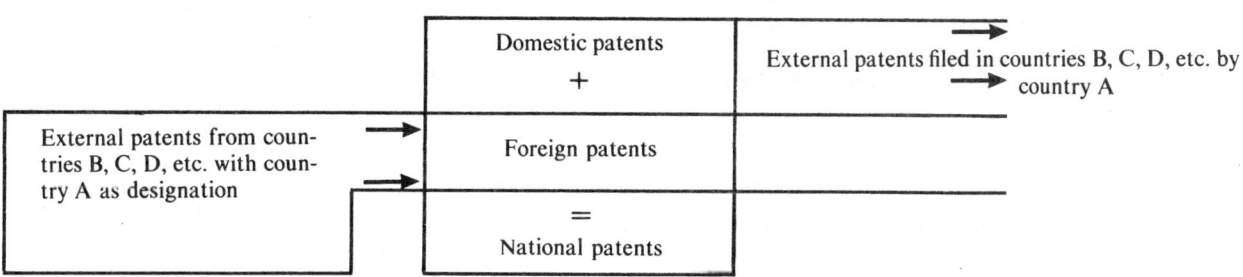

Overall Trends

One striking feature of patent applications in the OECD countries (see Graph 1 in Annex) is the end of growth and even a decline from the early seventies until 1978 in national patent applications, followed by a recovery mainly attributable to the stimulating effect of international patent channels on international patenting. Despite this general recovery, all major OECD countries, except Japan, showed very little growth during 1965-1983 and in Canada, Spain and Turkey, the absolute number of national applications declined. In several medium-sized and smaller countries, on the other hand, patent markets grew considerably.

Japan was the only major country with a continuous increase in national patent applications which tripled during the period under review. The growth in Japan is also clearly reflected by the trend in domestic applications (and grants) per capita of population (Table 28 in Annex) which contrasts sharply with the general decline elsewhere.

It is difficult to assess the significance of this general uptrend from 1978 onwards, since no information is available so far on the *use* of applications filed through international patent channels.

However, a number of concurrent trends can be identified:

- In most countries, the trend in national applications was mainly determined by foreign applications (including the United States but excluding Japan);
- The share of domestic applications in total national applications decreased in the *majority* of countries and only a minority of countries showed an absolute increase in domestic patenting activities;
- Trends in external patent applications followed that in total patents but showed more dynamic growth over the period. They were, however, low in absolute

terms except in the United States, Germany and Switzerland, where they were the most important patent category.

Lastly, the slowdown in domestic patent applications results in a fall in the apparent productivity of the resources allocated to the production of technology. In the OECD area as a whole, the number of business sector researchers per domestic patent increased from 1.8 in 1971 to over 3 in 1981. The exceptions to this trend were Australia, Ireland and Japan.

Position of the Countries as Domestic Markets

Regardless of different national definitions and methodological approaches, the situations of the various countries can be assessed in different ways. We shall consider the domestic aspect first, followed by the international one.

Confining our analysis mainly to the five or six countries whose patent markets account for at least a 5 per cent share of the OECD total (see Table 2.1), and which together represent 70 to 75 per cent of national applications in the OECD area, the most important features were:

i) Japan has experienced a marked uptrend and since 1968 has been responsible for the largest number of total national patent applications in the OECD area. By 1983, its share had reached 33 per cent (which corresponds to some 256 000 patents) or more than twice that of the United States' share of 13.5 per cent.

As Japan's share increased, those of the other major patent markets (Germany, United Kingdom, France) diminished. On the other hand, the Netherlands and Sweden had more growth in their patent markets and finished the period as they began. Austria and Finland were among the few countries where growth also occurred in relative terms. In all four countries this was due to trends in foreign patenting.

ii) Japan differs markedly from the other countries (see Graph 1 in Annex) as regards the breakdown of national applications between domestic and foreign. The rise in total applications in Japan is due, in fact, to the remarkable growth of domestic applications as can be seen when the growth rates of patent applications by type in Japan and in the United States are compared (see Table 2.2). The share of domestic patents in Japan rose from 74 to 89 per cent of national applications.

Table 2.1
National patent applications: country shares
OECD = 100

	1965	1970	1975	1980	1983
United States	17.5	16.8	17.2	15.7	13.5
Japan	15.1	21.3	27.2	28.6	32.7
Germany	12.3	10.8	10.2	9.8	9.3
France	8.8	7.7	6.9	6.6	6.3
United Kingdom	10.3	10.1	9.1	8.8	8.0
Italy	5.4	5.2	4.1	4.4	4.2
Canada	5.6	5.0	4.4	3.7	3.3
Spain	2.5	1.9	1.8	1.6	1.2
Australia	2.8	2.7	2.4	2.2	2.3
Netherlands	3.2	3.1	2.6	3.1	3.3
Turkey	0.1	0.1	0.1	0.1	..
Sweden	3.2	2.9	2.5	3.1	3.2
Belgium	3.1	2.8	2.2	2.4	2.7
Switzerland	3.4	3.2	2.9	3.1	3.1
Austria	2.2	1.9	1.7	2.3	2.5
Yugoslavia	0.4	0.5	0.6	0.5	0.3
Denmark	1.2	1.1	1.0	1.0	1.0
Norway	0.9	0.8	0.8	0.7	0.8
Greece	0.4	0.4	0.5	0.4	0.4
Finland	0.6	0.6	0.6	0.6	0.8
Portugal	0.2	0.3	0.3	0.3	0.2
New Zealand	0.6	0.6	0.6	0.5	0.5
Ireland	0.3	0.3	0.5	0.4	0.4
Iceland	0.0	0.0	0.0	0.0	0.0

Table 2.2
Growth in patent markets
1965 = 100

	1965	1970	1975	1980	1982
Japan					
National patents	100	160	195	237	313
– Domestic	100	165	222	273	375
– Foreign	100	144	117	133	136
United States					
National patents	100	109	107	112	112
– Domestic	100	105	89	86	82
– Foreign	100	121	164	198	210

Conversely, the increase in national applications in the United States reflects that of foreign applications which rose the fastest there. In consequence, the share of domestic applications fell from 76 to 56 per cent of the national total.

However, the United States is still one of the three countries (along with Japan and Germany) where domestic applications exceed 50 per cent of the national market. The average for the OECD area was 49 per cent at the end of the period under consideration, due to the weight of Japan.

iii) The Japanese performance is slightly less surprising if the nature of the applications filed and the proportion that result in the grant of a patent are taken into consideration.

First, until very recently, Japan had "single claim" patents[8] whereas in other countries each patent

Table 2.3
Patent grants: country shares
OECD = 100

	1965	1970	1975	1980	1982
United States	20.2	19.8	24.2	22.9	21.8
Japan	8.7	9.5	15.8	17.1	19.0
Germany	5.4	4.0	6.2	7.5	6.1
France	13.5	8.1	4.8	10.4	9.0
United Kingdom	10.9	12.6	13.8	8.8	11.1
Italy	6.5	9.1	..	3.0	2.1
Canada	7.8	9.0	6.9	8.9	8.4
Spain	3.5	2.3	3.1	3.4	3.7
Australia	2.3	1.9	4.1	3.1	2.2
Netherlands	0.8	0.8	1.3	1.2	2.5
Turkey	0.2	0.1	0.2	0.2	0.1
Sweden	2.6	4.2	3.1	1.9	3.0
Belgium	5.4	5.3	4.4	2.2	1.4
Switzerland	6.1	5.4	4.6	2.2	3.6
Austria	2.2	2.7	2.4	2.2	1.5
Yugoslavia	0.2	0.2	0.2	0.2	0.1
Denmark	0.9	1.0	0.8	0.6	0.6
Norway	0.8	0.8	0.7	0.8	0.7
Greece	0.7	0.9	0.6	0.8	0.8
Finland	0.3	0.4	0.5	0.7	0.8
Portugal	0.3	0.6	1.0	0.9	0.5
New Zealand	0.7	1.0	0.8	0.5	0.5
Ireland	0.2	0.2	0.3	0.5	0.4
Iceland	0.0	0.0	0.0	0.0	0.0
Total	100.0	100.0	100.0	100.0	100.0

Table 2.4
Shares in R&D and domestic patent applications in 1981

	Business enterprise R&D expenditure	Domestic patent applications
United States	50	18
Japan	15	54
EEC	28	21
Other	7	7
OECD	100	100

The EEC countries and the group of other OECD countries, on the other hand, have corresponding shares for both indicators in the OECD total.

Several of the other countries have a significantly higher share in domestic patenting activities than in R&D, notably Australia and New Zealand as well as Finland and Austria.

In Graph 18, the country situations in 1981 are shown for the same indicators (i.e. domestic patent applications and business enterprise R&D expenditure) but compared with the Domestic Product of Industry (DPI). This presentation has two advantages:

contains several claims. This means, for example, that an invention might need to be protected by several patents in Japan, which elsewhere might require only one. The number of domestic applications in Japan was thus inflated in relation to other countries.

Second, the use of data on patents actually granted casts a slightly different light on the trends in the shares of the major countries (see Table 2.3). The United States has consistently granted the largest number of patents with Japan in second place. Nevertheless, the uptrend in Japan is still very marked, since its share more than doubled between 1965 and 1982.

The use of data on patents granted also makes for some changes in ranking for the other countries, with France, the United Kingdom and Canada better placed than Germany.

Countries positions compared for R&D and patents

Relating business enterprise R&D expenditure to domestic patent applications underlines the findings for the United States and Japan in the previous section in that (as can be seen from Table 2.4) domestic patent activities are in inverse proportion to their R&D effort.

Graph 18
Patent intensity and R&D intensity*, 1981

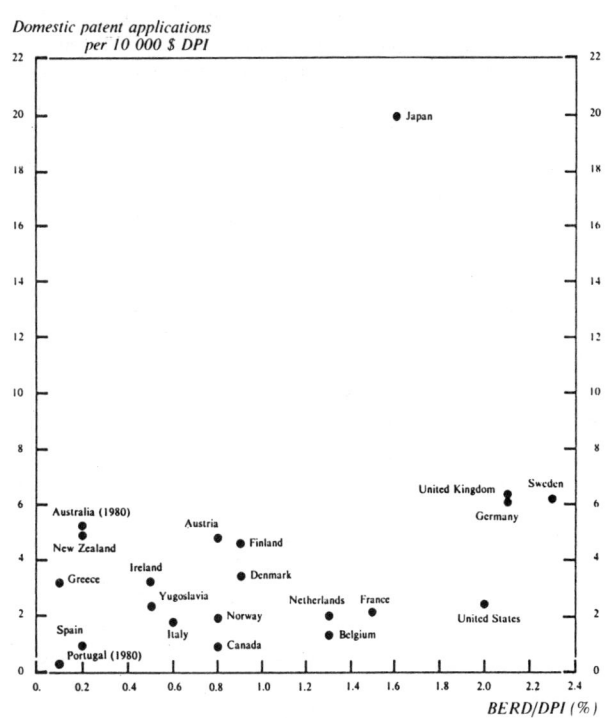

* Domestic patent applications and R & D expenditure in the business enterprise sector related to the gross domestic product in industries.
Source: OECD/STIIU Data Bank, November 1985.

51

i) the distortion caused by the extreme situations of the United States and Japan is eliminated;

ii) both patenting and research effort are expressed as "intensity" of the DPI.

The results can be summarised as follows:

- Although Japan still generates far more domestic patent applications per $10 000 DPI than any other OECD country, the three countries with the highest research intensity (Sweden, Germany and the United Kingdom) also have the highest relative patent output (after Japan, of course). However, the highly R&D intensive United States is level in relative patenting with less R&D intensive countries such as France, the Netherlands or Norway.
- On the other hand, in some countries where firms perform relatively little R&D, they do have a high propensity to patent given the size of their industrial economies. This was particularly true for Australia and New Zealand (as already mentioned earlier) but also for Austria and Finland. Canada, which is as R&D intensive as Austria, is five times less intensive in domestic patent applications.

POSITION OF THE COUNTRIES IN THE INTERNATIONALISATION OF PATENTS

The information obtained from international data on patents does not relate solely to the "production" of technology by the various countries. The data on both foreign and external applications can also be used to study the situation of each country as a centre for the dissemination of technology.

Two factors will be examined:

- the trend of external patent applications;
- the impact of the European patent and the PCT.

External applications

It is a widely held that, owing to the cost and time involved, patent applications filed abroad relate by and large to more "worthwhile" inventions than domestic applications[9]. The breakdown of such external applications among OECD countries and the relevant trends therefore supplement the earlier analysis of countries as producers of technology. From this point of view, the data in Table 2.5 also serve to put what might be regarded as Japan's "irresistible" rise into context.

If the countries are considered as sources of new technology in international circulation, they are still ranked in roughly the same order. The United States easily holds first place, despite a declining share, followed by Germany, whose situation had been steady over time but fell back in 1983. Japan's share has increased considerably from 3 per cent in 1965 to 13 per cent in 1983, behind Germany. Japan was the only major country (i.e. those with over 5 per cent of the total of external applications in the OECD area) to step up the absolute number of its external applications sixfold. France recuperated in 1983 the third place it held in 1965 through the use of international patent channels, whereas the United Kingdom dropped from third to fifth position. Both Finland and Australia drastically increased dissemination of their technology abroad, especially after 1978 and, with second highest growth after Japan, more than doubled their shares of the OECD total.

To sum up, the vast majority of countries accelerated the international dissemination of their technology during the last five years under review (see Graph 1 in Annex). Only a few countries showed stagnation or a little growth and two countries showed a decline (Portugal and Yugoslavia).

Table 2.5
External patent applications: country shares
OECD = 100

	1965	1970	1975	1980	1983
United States	36.9	34.6	31.0	29.8	31.7
Japan	3.0	7.4	9.2	11.6	12.9
Germany	18.9	19.6	20.2	21.1	17.9
France	7.0	6.8	7.8	8.4	8.0
United Kingdom	11.6	9.4	8.1	7.2	7.9
Italy	2.6	2.9	3.4	3.2	3.2
Canada	1.7	1.4	1.7	1.2	1.3
Spain	0.3	0.4	0.6	0.4	0.4
Australia	0.5	0.5	0.7	0.9	1.2
Netherlands	4.0	3.4	3.3	3.1	3.0
Turkey	..	..	..	..	..
Sweden	2.7	2.4	3.1	2.9	3.2
Belgium	1.4	1.1	1.1	1.0	0.9
Switzerland	6.9	7.3	6.6	5.8	5.1
Austria	1.0	1.1	1.1	1.2	1.1
Yugoslavia	0.1	..	0.1	0.1	0.0
Denmark	0.8	0.7	0.8	0.7	0.9
Norway	0.3	0.3	0.4	0.3	0.4
Greece	0.0	..	0.1	0.0	0.0
Finland	0.2	0.3	0.4	0.5	0.8
Portugal	0.0	0.0	0.2	0.0	0.0
New Zealand	0.1	0.1	0.2	0.2	0.2
Ireland	0.1	0.1	0.1	0.1	0.1
Iceland	..	..	..	..	..
Total	100.0	100.0	100.0	100.0	100.0

The impact of international filing procedures on dissemination of patents

The introduction of the European patent system (under the European Patent Convention, Munich, 5th October 1973) and an international system (under the International Patent Co-operation Treaty – PCT – of

19th June 1970) in 1978 have changed the behaviour of those filing patents abroad since it is cheaper to use these new systems to obtain wider international protection.

The introduction of new systems disrupted the national data compiled by the World Intellectual Property Organisation (WIPO) by increasing foreign and external applications through international filing. The resulting breaks in national series make it difficult to identify trends. On the other hand, the international systems have improved the comparability of the data as they introduce a de facto international harmonization in terms of both costs and examination procedures.

The influence of international patent channels on foreign and external patent applications can be illustrated by comparing external patent applications with foreign patent applications with and without EPC and PCT channels for each country. To do this, the number of external patent applications has been divided by the number of foreign patent applications separately for WIPO data only and for WIPO data supplemented by EPC and PCT figures. The ratios obtained were then compared to find out in which direction change occurred (see Table 2.6).

The results of this comparison are as follows: using national channels, in ten of the OECD countries more patents were "exported" than "imported" with Switzerland, Germany and the Netherlands in pole situation with external patent applications, twice to three times higher than foreign applications. The inclusion of international channels changed this pattern considerably: in only four countries (the United States, Japan, Germany and Switzerland) did external patent applications exceed the number of foreign applications. Column (c) illustrates the importance and direction of this change: the non-contracting countries were the only ones where more patent applications were "exported" than "imported" in relative terms as a consequence of international channels, indicated by the increase in the external/foreign patent applications ratio. Only Yugoslavia's ratio remained unchanged. These countries profited from the international channels by broadening their patent protection abroad without experiencing a significant increase in foreign patent applications at home. In all other (contracting) countries, the additional international channel led to a deterioration of the external/foreign patent balance although to varying degrees:

Table 2.6

Effect of EPC and PCT patents on external and foreign patent applications in 1983

Change in external/foreign patents ratios

	External patent applications/ Foreign patents applications		
	(a) = WIPO	(b) = WIPO + EPC + PCT	(c) = (b)/(a)
United States	1.54	2.89	1.88
Japan	1.41	1.92	1.36
Germany[1]	2.26	1.86	0.82
France[1]	1.45	0.90	0.62
United Kingdom	1.11	0.78	0.70
Italy[1] (1980)	1.02	0.53	0.52
Canada	0.15	0.24	1.60
Spain	0.14	0.20	1.43
Australia	0.24	0.44	1.83
Netherlands[1]	2.41	0.53	0.22
Sweden[1]	1.89	0.64	0.34
Belgium[1]	0.77	0.19	0.25
Switzerland[1]	3.58	1.08	0.30
Austria[1]	1.04	0.28	0.27
Yugoslavia	0.16	0.16	1.00
Denmark	0.30	0.60	2.00
Norway	0.17	0.27	1.59
Greece	0.03	0.06	2.00
Finland	0.66	0.74	1.12
Portugal	..	0.01	..
New Zealand	0.19	0.30	1.58
Ireland	0.07	0.15	2.14
Iceland	..	0.11	..

1. EPC signatory countries.
Source: OECD/STIIU Data Bank.

- The major signatory countries: Germany, France, the United Kingdom and Italy. Here, there was a simultaneous increase in both foreign and external applications, although less marked for the latter. The international systems thus result in increased penetration of patents of external origin.
- The smaller signatory countries: Austria, Belgium, the Netherlands, Sweden, Switzerland. Here, the impact was very marked as regards foreign patents, but much less so for external applications. The effect of the international systems in terms of providing an opening to foreign technology would, therefore, seem to be the stronger the smaller the country.

THE TECHNOLOGICAL BALANCE OF PAYMENTS AS A MEASURE OF THE INTERNATIONAL DISSEMINATION OF TECHNOLOGY

At first sight, there are two main routes for the international dissemination of technology: the transfer of technological information as measured by the TBP and the sale of goods incorporating technology, as discussed in the next chapter. However, a firm with technological assets has a third option: instead of selling licences or exporting, it can set up a subsidiary or make a direct investment. Moreover, once the parent firm has a network of subsidiaries, the conditions governing the choice between exporting, granting a licence or transmitting innovations through the subsidiaries are changed. Recent studies show that firms generally opt for dissemination through their subsidiaries.

This preference has an impact on international flows, as these are usually measured in terms of national frontiers: relocated production takes the place of exports and may give rise to the payment of royalties. This third route is all the more significant in that firms are more likely to be or to become multinational in industries where R&D and technological competition are decisive.

Characteristics of Technological Balance of Payments Data

Receipts and payments concern operations such as the transfer of patents, licensing agreements, provision of know-how, technical assistance, etc.

The two main problems as regards the recording of flows are: heterogenous contents and non-comparability at international level.

TBPs have a heterogenous content in that they record, side by side, not only flows relating to the transfer of technology proper (patents, manufacturing licences, know-how), but also in some countries services of a technical nature (assistance, training, consultancy work) and, in other countries, sometimes even factors related to industrial and intellectual property with no direct relationship to technology (trademark licences, film rights, management services, etc.).

The non-comparability at international level stems not only from the differences in coverage noted above, but also from variations in the survey procedures (direct/indirect with regard to enterprises, exhaustive or on a sample basis) and in the way the information is presented (broken down by type of firm, by activities, etc.).

Problems of interpretation are raised not only by the mixed contents of the TBPs, but also by the elusive character of certain international flows of technological knowledge, i.e. those for which there is no visible form of payment (among others: cross-licensing, transfer of knowledge to a subsidiary, international co-operation of a non-commercial type). Another such factor is the behaviour of the firms mainly responsible for transferring technology: i.e. multinational enterprises, for whom the type of payments registered in the TBP may be only one of several possible channels of reimbursement for technology transferred to subsidiaries. Their choice between these channels will be affected by fiscal and other considerations which may lead to the TBP data seriously overestimating or underestimating the real flows of technology involved. This means it is difficult to make an economic interpretation of intra-firm flows on the basis of accounts which, in the last analysis, are based on the firms' worldwide strategy.

Overall Trends

At present, OECD has collected data on the TBP for only just over half of the OECD countries and a complete set of figures for the period 1972-82 is, as yet, available for only nine[10] of them. However, the latter represent 80 to 90 per cent of the available total, as they include the six most important countries for payments and receipts. It can be safely assumed that the inclusion of the missing OECD countries would not change the following general remarks significantly.

The first general phenomenon to be noted is the concentration of receipts and payments in a limited number of countries.

The second concerns the trend in receipts and expenditure of the nine countries grouped in Graph 19, where the relative stability of payments contrasts with the rather irregular increase in receipts until 1980 and its subsequent decline.

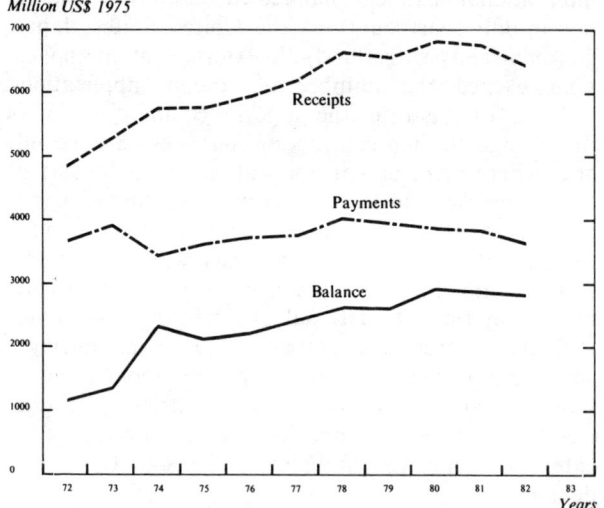

Graph 19
Technological balance of payments: Nine countries total

Source: OECD/STIIU Data Bank, November 1985.

The contrasting trends in receipts and payments can be read as suggesting that there has been a relative increase in transfers of technology to industrialising countries. The reasoning is as follows: as OECD counts payments which are almost exclusively made to (and thus received by) other OECD countries, the more rapid growth in receipts must come from non-OECD countries mainly and, it is assumed, from developing ones.

Position of the Countries with regard to the International Dissemination of Technology

i) Trends, levels and structure of technology transfer

The ranking of the countries under consideration and the relevant trends are shown in Table 2.7 in terms of receipts and payments, balances and coverage ratios.

The first point of note is the exceptional situation of the United States' economy as regards the balance, the coverage ratio and its share in receipts. The United States remains undoubtedly the principle centre for the dissemination of technology.

The United Kingdom was the only other major OECD country with a positive (and growing) technology balance though this was lower in absolute and relative (coverage ratio) terms. The only others with positive balances were Denmark and Sweden.

As already noted above, receipts in most countries grew more rapidly than payments, leading to declining deficits (or growing balances) and improved coverage ratios. This phenomenon was particularly marked in Japan, the only country where receipts tripled at fixed prices and payments fell. By 1983, the Japanese deficit had fallen to less than one-third of those of Germany and Italy[11].

Table 2.7 confirms that the dissemination of technology is highly concentrated around the United States. The most significant changes on the receipts side were the declining shares of the United States and the United Kingdom and the growth in those of Japan, France and Spain. Payments are less concentrated, with Japan and Germany spending about the same amounts in the purchase of technology, followed by France, Spain, the United Kingdom and Italy which, despite their varying economic size, also spend heavily on foreign technology in absolute terms.

To summarize the situations of the various countries in terms of balances or coverage ratios over the period (at fixed prices):

- in Japan, the United States and the United Kingdom, coverage ratios have improved due to increased receipts *and* falling payments, the trend in Japan being towards reduced dependence and a larger role in dissemination;
- an improved coverage ratio is associated with either a stagnation in flows (United Kingdom) or increased flows (France);
- slight improvement in the coverage ratio due to stronger increases in receipts than in payments (Austria, Finland, Spain, Italy and Germany);
- the Netherlands and Portugal had a worsening coverage ratio owing to a more marked increase in payments.

ii) Technology Transfers compared with Industrial R&D Efforts

The scale of the various countries' technological payments and, more especially, of their receipts are related to their size and technological potentials. Without trying to establish complex correlations that would call for very lengthy analysis, a relationship can be said to exist between receipts and payments and R&D expenditure, but it is not simply one-to-one.

The purchase of foreign technology, and the related payments can be treated as equivalent to an import of

Table 2.7

Technological balance of payments in OECD countries

In million US$ at 1975 prices and as percentage

	1973						1983					
	Receipts	%	Payments	%	Balance	Coverage ratio	Receipts	%	Payments	%	Balance	Coverage ratio
United States	3 582.6	(65.7)	456.0	(9.8)	+3 126.0	7.85	4 328.6	(59.4)	132.2	(2.8)	+4 196.4	32.74
Japan	231.2	(4.2)	1 035.7	(22.2)	−804.5	0.22	645.9	(8.9)	748.9	(16.0)	−102.9	0.86
Germany	222.8	(4.1)	618.4	(13.3)	−395.5	0.36	365.1	(5.0)	728.7	(15.6)	−363.6	0.50
France	328.3	(6.0)	476.5	(10.2)	−148.2	0.69	557.7	(7.6)	617.8	(13.2)	−60.2	0.90
United Kingdom	647.2	(11.9)	552.7	(11.9)	+94.5	1.17	650.7	(8.9)	5 140.4	(10.9)	+140.3	1.27
Canada	45.3	(0.8)	188.7	(4.0)	−143.4	0.24	122.1	(1.7)	260.2	(5.6)	−138.1	0.47
Italy	92.7	(1.7)	469.2	(10.1)	−376.5	0.20	119.8	(1.6)	485.8	(10.4)	−366.0	0.25
Netherlands[1]	165.7	(3.0)	222.1	(4.8)	−56.4	0.75	209.1	(2.9)	351.3	(7.5)	−142.2	0.60
Sweden	23.9	(0.4)	21.8	(0.5)	+2.0	1.09	58.1	(0.8)	26.6	(0.6)	+31.5	2.19
Austria	10.8	(0.2)	58.4	(1.3)	−47.7	0.18	16.9	(0.2)	88.4	(1.9)	−71.6	0.19
Denmark[2]	48.7	(0.9)	39.2	(0.8)	+9.5	1.24	59.9	(0.8)	40.0	(0.9)	+19.9	1.50
Finland[1]	3.7	(0.1)	33.6	(0.7)	−29.9	0.11	28.6	(0.4)	44.6	(1.0)	−16.0	0.64
Portugal	2.7	(0.0)	25.6	(0.5)	−22.7	0.11	4.8	(0.1)	47.7	(1.0)	−42.9	0.10
Spain	51.0	(0.9)	462.2	(9.9)	−411.2	0.11	123.4	(1.7)	596.9	(12.8)	−473.5	0.21
Total	5 456.6	(100.0)	4 660.6	(100.0)	n.c.	1.17	7 290.7	(100.0)	4 679.5	(100.0)	n.c.	n.c.

1. 1982 instead of 1983.
2. 1981 instead of 1983.
n.c.: Non-cumulative.
Notes: Excluding Australia because of lack of data.

technology which supplements national supply. The ratio of technological payments to the R&D expenditure of the Business Enterprise sector may serve as a yardstick of dependence on external sources of technology (always allowing for the fact that it will also reflect the degree of foreign ownership in a given economy).

Similarly, if technological receipts are assimilated to an export of technology, the ratio of receipts to business enterprise R&D provides a rough estimate of a country's apparent technological competitiveness and/or its successful application on foreign markets, e.g. through direct investment.

These calculations also adjust the data to take into account the large differences between countries in the size of their national industrial sectors and thus their technological potentials which, in turn, largely reflect their national economies.

Tables 2.8 and 2.9 show the values of the two ratios for the years 1971 and 1981 for OECD countries for which data are available. In terms of trends over time, they reveal that, in virtually all Member countries, home R&D spending grew more rapidly than payments for foreign technology. The trend for receipts, however, varied between countries when compared with industrial R&D.

The tables also reveal that, although the five major countries are responsible for the bulk of flows measured by the technological balance of payments, international purchase and sale of technology play a relatively small role compared with home R&D efforts (less than 20 per cent of BERD). This is also true in Sweden at much lower absolute levels. The decline in the purchase of technology compared with home industrial R&D was particularly marked in Japan, whereas the growth in receipts was notable in France where, by 1981, revenue from the sale of technology had reached the same percentage of BERD as in the United Kingdom and the United States. Three of the four countries with positive

Table 2.8
Technological payments as percentage of business enterprise R&D

	1971		1981	
United States	1.3		1.3	
Japan	15.0		7.2	
Germany	14.1		10.2	
France	15.8		14.6	
United Kingdom	14.8	(1972)	10.5	
Australia	..		36.4	
Canada	23.7		24.6	
Italy	41.3	(1972)	28.3	
Netherlands	26.3		41.9	
Sweden	4.3		3.8	
Austria	40.0	(1972)	23.0	
Denmark	33.6	(1970)	22.8	
Finland	34.1		26.4	
Portugal	136.7	(1972)	154.1	(1982)
Spain	235.2	(1973)	158.2	

Table 2.9
Technological receipts as percentage of business enterprise R&D

	1971		1981	
United States	13.0		13.2	
Japan	3.0		4.8	
Germany	5.2		4.7	
France	8.5		13.4	
United Kingdom	16.3	(1972)	12.7	
Australia	..		3.6	
Canada	6.7		7.3	
Italy	7.1	(1972)	9.9	
Netherlands	23.9		27.3	
Sweden	4.8		4.0	
Austria	7.5	(1972)	5.5	
Denmark	38.6	(1973)	34.2	
Finland	4.0		16.9	
Portugal	11.2	(1972)	25.2	(1982)
Spain	26.0	(1973)	50.4	

TBP balances were in this group (Sweden, the United Kingdom, the United States).

A second group of countries where technological flows came to 20-40 per cent of home business enterprise R&D comprises Italy, Canada, Australia and the majority of small Member countries. The Netherlands is the major exception to the trend of declining purchases of technology compared to home R&D, partly explained by slow growth in R&D spending. Denmark is the only country in the group with a positive TBP balance.

A third group includes two countries – Portugal and Spain – whose spending on foreign technology exceeds their own R&D efforts owing to their levels of industrial development and the weakness of their technological potentials.

iii) *Types of Payments and of Firms*

In addition to the aggregated data analysed above, it would be useful to have some more detailed information concerning the types and contents of apparent technological dependence/competitiveness, especially concerning two structural factors. The first is the composition of the technological receipts and payments. It was pointed out above that the balances are not homogeneous and that technical assistance, studies, etc. are often included as well as the patents and licences corresponding to technological transfers in the strict sense. An analysis of the balances of the various countries, making a distinction at least between patents and licences on the one hand and technical assistance or management fees on the other, can provide useful additional information on the question of apparent dependence/competitiveness.

The case of France illustrates the above point[12]. From 1970 to 1982, the TBP showed a steady overall negative balance at current prices, and a decreasingly negative one in purchasing power parities. However, this relative improvement masks a substantial deficit as regards patents and licences, which was offset by the increase in receipts from management fees.

The second structural factor concerns the participation of multinational enterprises in international flows of technology and the related payments. Few countries provide a breakdown by type of ownership of firms. The OECD has, as yet, only been able to compile relevant data for the United States and the United Kingdom and to some extent for Germany[13]. Analysis of the trends over fifteen years (1967-1983) shows that in the United States and the United Kingdom, the share of receipts earned by the subsidiaries of multinational companies grew significantly (see Table 32 in Annex).

In the United Kingdom, their share in total receipts increased from a third to over one half and in the United States from three-quarters to over four-fifths. This concentration in technology transfer linked to direct investment underlines the thesis that companies prefer to transfer technology abroad via their subsidiaries giving them more control over its use and more protection against competitors.

This is reflected in the payments made by subsidiaries of foreign companies, e.g. in Germany. On average, three-quarters of all payments were made by such firms and they are the principal cause of the German deficit (the balance for "enterprises in which there is no major foreign interest" has always been positive and has improved over time). The proportion of payments by affiliated companies in total payments increased noticeably in the two other countries, reaching almost 90 per cent in the United Kingdom and 80 per cent in the United States.

The British balance for related firms shows a deficit which is tending to diminish and which is largely compensated for by a surplus for unrelated firms, whereas for the United States both balances were positive over time with the balance for related firms generating the high surplus.

This is all that can be said at the present moment on the basis of TBP series currently stocked at OECD because of the lack of data and, more seriously, of international standards for their collection and use as science and technology indicators. With the help of Member countries, OECD hopes to take positive steps to deal with these problems so that more reliable analyses of the diffusion of technology can be made in subsequent reports in this series.

However, when trying to pull together research, patenting and technology flows for the three major geographical and economic OECD areas, some tentative conclusions may be drawn, highlighting the different strategies of their respective industries and which set the scene for the following chapter on trade in high technology industries.

Table 2.10
Shares in 1981 of selected OECD countries

	BERD	External pat. appl.	TBP receipts
United States	54	36	65
Japan	16	14	7
EEC[1]	30	50	28
Total OECD	100	100	100
	BERD	Foreign pat. appl.	TBP payments
United States	54	21	12
Japan	16	13	19
EEC[1]	30	66	69
Total OECD	100	100	100

1. Germany, France, United Kingdom, Italy, the Netherlands and Denmark.

Table 2.10 indicates that, using industrial R&D as a yardstick, the United States clearly preferred to transfer technology abroad – mainly via subsidiaries – whereas the reverse is true for Japan.

The six countries of the EEC area, together, spend over two-thirds of the total for the purchase of foreign technology which indicates their relatively high dependence on foreign multinationals.

It appears that both foreign and European inventors see the EEC area as an interesting market for technological applications as reflected by the high share of foreign patents in Europe. On the other hand, filing of external patent applications series suggests that European countries are also preparing to penetrate foreign markets.

Chapter II

TECHNOLOGICAL PERFORMANCE AND INDUSTRIAL COMPETITIVENESS

The analysis in Chapter I was largely based on indicators designed to measure output in terms of science and technology. Country positions and relative trends were presented purely within the framework of the production and international dissemination of technology. The exercise will now be broadened by relating technological performance to industrial competitiveness. This relationship, which may seem obvious, is nonetheless difficult to grasp, for it works both ways.

It is reasonable to assume that a country's industrial competitiveness – or its ability to compete – partly depends on the quality of the products and processes used. This dependence is all the greater in industries where the competitors can copy new technically sophisticated products (high R&D intensity industries), but it also plays a less direct role in other industries, including those considered to be traditional. For example, the textile or steel industries in developed countries can stand up to the new competitors who have moved into the international market only by acquiring high-performance equipment that owes a great deal to advances in electronics and computers. This one example is sufficient to show the importance of this interdependence between industries, especially via the supply and purchase of capital goods, as this is, in fact, the way the effects of technical progress are disseminated.

Although industrial competitiveness depends on the effectiveness of technological thrust, it can also be shown that technological performance in its turn is partly determined by general factors relating to industrial competitiveness. In other words, measuring technological performance in terms of the exports of high R&D intensity industries involves measuring, at one and the same time, efficiency in the development of new technologies in their industrial application and in the marketing of the resulting industrial products. What is measured in every case is the result of a set of activities carried out by enterprises which require a series of skills to be carried out successfully.

But what was already true for the technology dissemination indicator based on exports with a high technological content is also true of external patent applications and of earning royalties from the international use of patents and know-how.

On the one hand, there are several alternative or complementary ways in which firms can exploit new technologies at international level – exports, subsidiaries or selling licences to independent firms.

On the other hand, there is a marked interdependence between certain individual indicators (for example, between an external patent application and direct or indirect exploitation of the market respectively through exports and licences and/or subsidiaries, and the interdependence between the multinationalisation of firms and the payment of royalties).

The trend in the competitiveness of the various industries (with a high, medium and low R&D intensity) will be analysed in order to study the two-way relationship between technological performance and industrial competitiveness. But, first, we shall endeavour to throw light on the relationships between the production of new technology and changes in industrial structures measured in terms of trade by high R&D intensity industries.

IDENTIFYING HIGH, MEDIUM AND LOW R&D INTENSITY INDUSTRIES

The "high-tec" concept

One way in which technology is diffused internationally is incorporated in goods. It would seem plausible that products with the highest new technology content would come from industries with the most substantial R&D efforts. Furthermore, the degree to which a country is successful in exporting such products could be used as an indicator of its international technological position as well as of its general competitiveness[14].

Although these ideas seem attractive at first sight, difficulties arise as soon as we start to try and define exactly which activities are to be analysed.

First, there are no clearly agreed concepts but rather a multiplicity of rather similar terms, such as "high technology", "advanced technology", "core technology", or "strategic technology", all of which are based on the idea that R&D and technological know-how contribute to industrial and commercial success.

The following characteristics are usually attributed to activities considered as being in the high-technology class:

- Need for a strong R&D effort;
- Strategic importance for governments;
- Very rapid product and process obsolescence;
- High-risk and large capital investments;
- High degree of international co-operation and competition in R&D production and worldwide marketing.

Defined in terms of R&D intensity

Manufacturing industries could, in theory, be classified using all the above criteria. However, the Secretariat's work so far has been based exclusively on the first of these. In addition to the fact that they have not yet been quantified, the other characteristics pose certain methodological problems, in that industries cannot be classified in a way which satisfies all the criteria simultaneously. In consequence, at this stage in the work, only the first criterion has been used. High technology industries are, thus, those with a high R&D intensity.

This report uses the approach by industry rather than that by product as adopted in an earlier study. The level of aggregation is that used in the OECD's international surveys on personnel and expenditure devoted to R&D[15].

Technological or R&D intensity is measured by the ratio of R&D expenditure to production. This ratio is calculated for each industry for the eleven reference countries taken together as an area[16]. It is an average weighted by each industry's share in total output for the eleven countries.

The industry ranking for the years 1970 and 1980 is given in Table 2.11. Six industries are in the high intensity category with a growing average coefficient due to the computer and electronics industries.

In the process of defining high R&D intensity industries, two other categories of industrial activities

Table 2.11

Intensity of R&D expenditure in the OECD area

Weighting of the 11 main countries – R&D expenditure/output

1970		1980	
	Intensities		Intensities
High		**High**	
1. Aerospace	25.6	1. Aerospace	22.7
2. Office machines, computers	13.4	2. Office machines, computers	17.5
3. Electronics & components	8.4	3. Electronics & components	10.4
4. Drugs	6.4	4. Drugs	8.7
5. Instruments	4.5	5. Instruments	4.8
6. Electric machinery	4.5	6. Electrical machinery	4.4
Average	10.4	Average	11.4
Medium		**Medium**	
7. Chemicals	3.0	7. Automobiles	2.7
8. Automobiles	2.5	8. Chemicals	2.3
9. Other manuf. ind.	1.6	9. Other manuf. ind.	1.8
10. Petroleum refineries	1.2	10. Non-electrical machinery	1.6
11. Non-electrical machinery	1.1	11. Rubber, plastics	1.2
12. Rubber, plastics	1.1	12. Non-ferrous metals	1.0
Average	1.7	Average	1.7
Low		**Low**	
13. Non-ferrous metals	0.8	13. Stone, clay, glass	0.9
14. Stone, clay, glass	0.7	14. Food, beverages, tobacco	0.8
15. Shipbuilding	0.7	15. Shipbuilding	0.6
16. Ferrous metals	0.5	16. Petrol refineries	0.6
17. Fabricated metal products	0.3	17. Ferrous metals	0.6
18. Wood, cork, furniture	0.2	18. Fabricated metal products	0.4
19. Food, beverages, tobacco	0.2	19. Paper, printing	0.3
20. Textiles, footwear, leather	0.2	20. Wood, cork, furniture	0.3
21. Paper, printing	0.1	21. Textiles, footwear, leather	0.2
Average	0.4	Average	0.5

have been identified, i.e. medium-intensity and low-intensity industries (see Table 2.11). The stability in the breakdown which was already observed in the first category is confirmed for the other two. Only the petroleum refining and non-ferrous metals industries change category.

Common characteristics

The specific characteristics of individual high R&D intensity industries seemed to call for more detailed examination, to see whether they have enough in common to differentiate them from the other two industry categories and thus warrant their being dealt with separately.

The following two-stage procedure was used: first, each industry, considered at OECD total level, was assigned a series of "characteristic values" based on nine "characteristic" indicators, each of which could have one of three "values" (low, medium and high) defined in terms of the indicator's average value over the 1970-1980 period.

The following indicators were selected:

– Weight of the industry in total manufacturing output;
– Weight of the industry in total manufacturing exports;
– Weight of the industry in total manufacturing imports;
– The industry's export ratio;
– The import/output ratio in each industry;

Graph 20
Factor analysis of trade profiles of manufacturing industries
Without technological variables

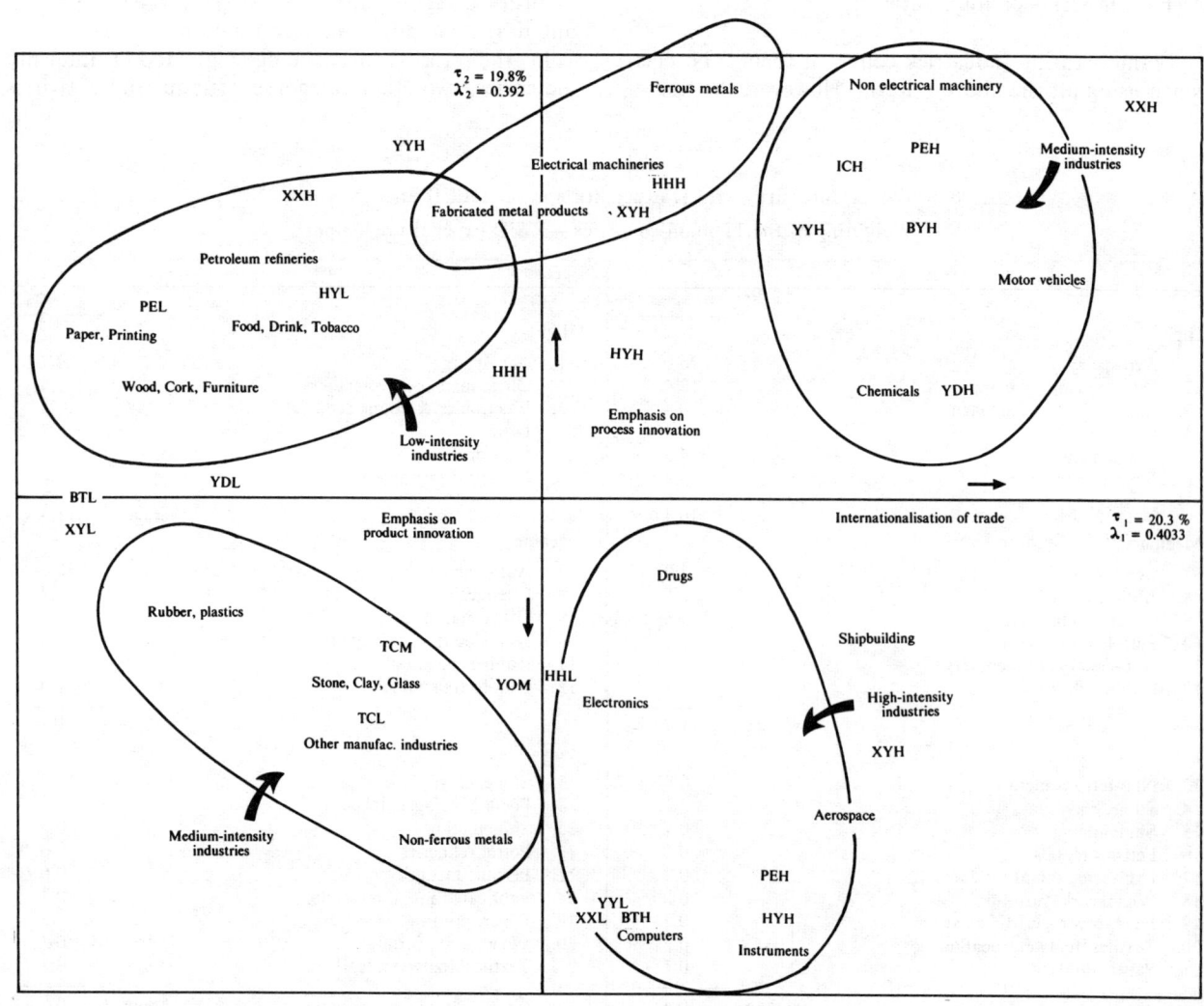

Source: OECD/STIIU Data Bank, November 1985.

- Trade balance in relation to overall average trade;
- Surplus output in relation to domestic demand (specialisation by "niche");
- Import penetration rate;
- Export/import ratio.

These indicator values were used to construct a matrix with the rows showing the 21 industries broken down into three categories, and the columns the 9 indicators broken down into three categories (Table 33 in Annex).

The second step was to conduct a factor analysis of correspondence, a statistical method for the purpose of ascertaining the factors explaining the resemblances between the rows and columns in such a matrix, with a minimum loss of information.

The results are shown in Graph 20. Two explanatory factors (represented by the two axes) together account for 40 per cent of the variance: one is the degree of internationalisation of trade, and the other the relative importance of process innovation and product innovation.

The projection of the information contained in the matrix, on to these two axes, reveals four categories of industries. The high R&D intensity industries emerge as a group in bottom right quandrant, clearly distanced from the group of low-intensity industries in top left quandrant. The medium-intensity industries, however, are projected in two opposed groups (top right quandrant and bottom left quandrant).

These positions can be interpreted as follows:
- The trade profiles of high R&D intensity industries are well correlated. They correspond to a high degree of *internationalisation of trade* (rate of exports and exposure to foreign competition through imports) associated with the predominance of *product innovation*. The profile of one industry in the category, electrical machinery, is nearer to that of the medium-intensity industries.
- The low-intensity industry profile combines a low degree of internationalisation with a tendency to process innovation, a characteristic of mature industries.
- The medium-intensity industries are divided into two sub-groups: the first combines a high or medium degree of internationalisation and process innovation (automobiles, chemicals, non-electrical machinery, electrical machinery, ferrous metals and fabricated metal products); the second combines product innovation and a low or medium degree of internationalisation (rubber, plastics, stone, clay, glass and non-ferrous metals).

The main conclusion of this analysis is that the high-intensity industries do have special characteristics enabling them to be considered together as a specific category[17].

R&D AND INDUSTRIAL STRUCTURES

General importance of the three groups of industries at all OECD level

Taken together, the R&D intensive industries are responsible for most of the industrial R&D expenditures in the area, accounting for 51 per cent of the total on average during the 1970-80 period. Their average share over the same period is, however, much lower in terms of production (11 per cent) or of exports (16 per cent) (see Table 2.12).

Before the specificity of the first category is examined in greater detail, it should be stressed that 89 per cent of output and 84 per cent of exports are from the second and third categories. In other words, the review of

Table 2.12

Average weights of the manufacturing industries of the OECD area (11 countries) during the period 1970-1980

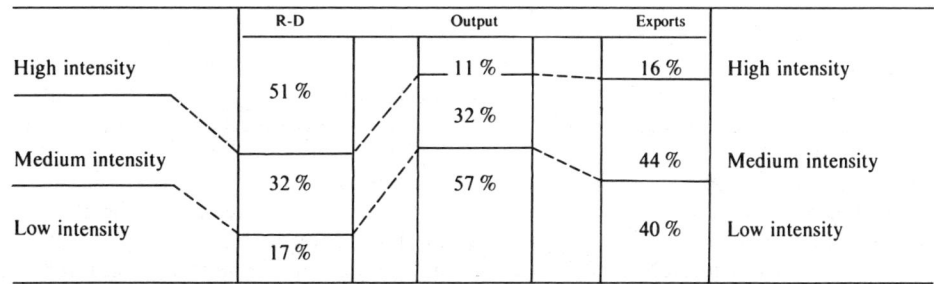

industrial performance obviously cannot be limited to the high R&D intensity category of industries.

This difference can be brought out more clearly if the relative propensities[18] of each category to export and conduct R&D work is measured.

Industry category	Relative propensity to conduct R&D	Relation propensity to export
– High intensity	4.64	1.45
– Medium intensity	1.00	1.36
– Low intensity	0.30	0.77

The differences appear greater in terms of R&D, but they are by no means negligible at export level.

More about high R&D intensity industries

The specific characteristics of high R&D intensity industries can be examined in terms of R&D itself and of exports.

R&D: concentration and public funding

Most industrial R&D is concentrated in the high R&D intensity industries especially in four major countries: the United States, the United Kingdom, France and Germany, where it is well over half of the total. It is lower in Japan where it was 40 per cent in 1981. This concentration relates to R&D expenditures, so the question naturally arises as to the role played by government support for R&D in a category of industries, some of which are unquestionably of strategic interest. Attention may be drawn to two aspects: the breakdown of government finance and its share in total R&D expenditure.

Government R&D finance is even more concentrated in high-intensity industries than is total R&D expenditure (see Table 2.13). Some 70 to 95 per cent of government money goes to the first category. The main exception is Japan where medium and low-intensity industries receive three-quarters of government R&D finance, although it should be stressed at once that government supports relatively little R&D in any of the categories.

Table 2.13

Proportions of public funding in total R&D in the high, medium and low R&D intensity industries

Approximate estimates

	1970			1980		
	High	Medium	Low	High	Medium	Low
United States[1]	56.0	13.0	4.0	42.0	10.0	13.0
Japan	0.8	0.7	1.0	0.6	0.3	6.0
Germany[2]	22.0	3.0	9.0	19.0	7.0	15.0
France	..	..	..	36.0	5.0	5.0
United Kingdom[3]	55.0	7.0	4.0	49.0	4.0	4.0
Italy[4]	7.0	2.0	2.0	5.0	2.0	6.0
Canada[5]	18.0	8.0	8.0	11.0	6.0	6.0
Sweden[6]	15.0	12.0	12.0	11.0	12.0	7.0

1. On the technical assumption that the allocation of public funding between the chemical and drugs industries on the one hand and the computer and non-electrical machine industries on the other is proportional to these industries' total R&D effort.
2. Secretariat estimate, on the assumption that 15 per cent of public R&D funding for non-electrical machines goes to the computer industry and that 30 per cent of funding for chemicals goes to the drugs industry.
3. It is assumed that 10 per cent of the funding for chemicals goes to the drugs industry. The latest year for which information is available is 1978.
4. On the assumption that 6 per cent of funding for transport goes to the aerospace industry.
5. Assuming that 85 per cent of funding for transport goes to the automobile industry and 53 per cent of funding for non-electrical machinery to the computer industry.
6. On the assumption that 30 per cent of the transport category is accounted for by aerospace and 50 per cent by the automobile industry, and that 16 per cent of funding in the machine category is accounted for by the computer industry.

This is shown by an estimate of what proportion of total R&D expenditure is financed by government in the three categories. In Japan this share received a maximum of 6 per cent in 1980 in the case of low-intensity industries (see Table 2.14).

Examination of the percentage of all R&D in high-intensity industries financed by governments shows that, despite a downtrend between 1970 and 1980, it still remained very significant in the United States, France and the United Kingdom, as this category accounted for respectively 42 per cent, 36 per cent and 49 per cent of total R&D expenditures in 1980. Though lower in other countries, this relative government contribution is still higher in the high-intensity industry category than in the others.

As has already been noted in Part I, a high relative level of government support for R&D in the high intensity group of industries seems to be associated with

Table 2.14[1]

Public funding for R&D in the high, medium and low-intensity industries

Approximate estimates

	1970			1980		
	High	Medium	Low	High	Medium	Low
United States	92.0	7.0	1.0	88.0	8.0	4.0
Japan	26.0	40.0	34.0	21.0	12.0	67.0
Germany	85.0	10.0	5.0	67.0	23.0	10.0
France	..	..	..	91.0	7.0	2.0
United Kingdom	92.0	6.0	2.0	95.0	3.0	2.0
Italy	76.0	21.0	3.0	69.0	18.0	13.0
Canada	73.0	15.0	12.0	67.0	16.0	17.0
Sweden	60.0	15.0	25.0	71.0	20.0	9.0

1. Subject to the same assumptions as for Table 2.13.

defence and aerospace programmes and the existence of a sizeable aerospace industry, notably in the United States, the United Kingdom and France (see Part I, Chapter II).

Trade of high-R&D intensity industries combined

There is a positive correlation between the share in manufacturing exports taken by high R&D intensity industries and the R&D intensity of these industries. This correlation was statistically significant when tested for the eleven countries for 1970 and 1980 (see Graph 2 in Annex).

A country's position in international trade by high R&D intensity industries can be studied from various viewpoints.

a) *Output and exports*

A comparison of the shares of high-intensity industries in industrial output and exports in each country already gives a number of pointers. The United States, thus, specialises the most in this type of exports. Its share has even increased, although the opposite occurred for output. The United States seems to hold a relative advantage in high intensity R&D industries, which is probably due to its big R&D efforts (63.6 per cent of the total, the highest proportion in the OECD area).

A second finding is that the share of high-intensity industries in exports increased faster than its share in output. This denotes greater emphasis on trade, consistent with the previous remarks on the propensity to export. With the exception of the Netherlands, the export share of output[19] has risen, particularly in the United States, the United Kingdom, Japan and Canada.

The third point to be noted is that in 1980 the weight of high-intensity industries in exports in the five major countries (United States, Japan, United Kingdom, France and Germany) was higher than the OECD average (16 per cent for the 1970-80 period). Detailed output data are available for the eleven main OECD countries only (see Table 2.15). For this reason other countries could not be included in any tables requiring output data.

Table 2.15

Weights of high R&D intensity industries in total manufacturing output and exports

	Output			Exports		
	1970	1975	1982	1970	1975	1982
United States	14.6	12.5	10.8	25.8	24.6	31.1
Japan	14.1	12.2	13.4	20.2	17.2	26.9
Germany	11.9	12.1	12.0	15.5	14.6	17.7
France	10.3	11.1	11.3	13.9	13.6	18.2
United Kingdom	12.2	11.2	12.5	16.8	18.8	24.8
Italy	11.7	11.2	12.1	11.5	9.8	12.3
Canada	8.5	7.5	6.7	8.8	7.7	10.0
Australia	7.2	8.3	7.6	2.7	4.4	3.7
Netherlands	11.8	12.3	12.3	15.9	14.1	12.8
Sweden	9.6	10.3	10.0	11.7	12.9	14.6
Belgium	6.5	6.8	6.6	7.1	8.5	8.8
EEC	11.4	11.3	11.7	15.9	15.5	16.8

b) *Balances, specialisation by "niche" and intra-industry specialisation*

The trade balances of the high-intensity industries as a whole are positive only in the five major countries and two small ones, Switzerland and Ireland. This result should be interpreted in the light of the size of the domestic market in these economies, for the biggest countries can specialise in several of these industries, while the smaller countries must restrict themselves to only one or two. Structurally speaking, it is therefore more difficult for the latter countries to achieve a trade balance in their high R&D intensity industries as a whole.

Switzerland and Ireland are exceptions but for different reasons. Swiss manufacturing industry spends heavily on R&D in relative terms (R&D expenditures/value added), almost as much as the United States. Strong Swiss specialisation in drugs and scientific instruments ensures large enough surpluses to cover the deficit for low R&D intensity industries. This strong position has been reinforced by additional exports resulting from attempts to invest abroad. The Irish surplus in trade by high intensity industries results from the existence of a large number of foreign firms, many in the electronics industry, which are responsible for 60 per cent of Irish industrial exports.

The information provided by the balances is supplemented by two specialisation indicators. They concern:

- Specialisation by "niche" (SN) which is defined as the ratio of output to domestic demand. The higher the value of this indicator, which measures the surplus of output over domestic demand, the greater is the country's specialisation (positive balance). Values of around 100 mean that specialisation is balanced.
- Intra-industry specialisation (IS) which is given by the ratio:

$$100 \frac{X - M}{X + M}$$

This ratio ranges from +100 (exclusively exporting country) to –100 (exclusively importing country). It takes into account the total volume of trade. A big surplus (X–M) is required for the IS ratio to be well above 0 for a country heavily involved in trading (substantial X+M).

Table 2.16 shows the trend in the values of these two indicators.

Table 2.16

Specialisation by "niche", intra-industry and apparent advantage in high R&D intensity industries

	S.N.			I.S.			C.A.		
	1970	1975	1980	1970	1980	1984	1970	1980	1984
United States	105	110	109	37	24	–3	158	156	156
Japan	109	114	126	43	61	70	123	141	147
EEC[1]	106	110	104	18	7	1	93	93	82
Germany	118	121	115	27	14	11	96	93	82
France	100	104	102	–1	1	5	85	81	83
United Kingdom	108	113	108	17	6	–4	104	121	118
Italy	99	99	94	3	–7	–5	77	62	56
Netherlands	93	100	97	–8	–4	–6	97	76	62
Belgium-Luxembourg	77	81	79	–20	–13	–10	44	47	37
Denmark	..	..	..0	–26	–8	–3	72	75	66
Ireland	..	..	..	–36	–1	13	71	113	159
Greece	..	..	..	–92	–75	–72	14	17	13
Canada	81	73	66	–27	–39	–38	55	48	43
Australia	61	66	64	–83	–71	–80	17	26	19
New Zealand	..	..	..	–91	–77	–75*	4	9	9*
Austria	..	..	..	–17	–19	–14	69	73	63
Finland	..	..	..	–62	–36	–34	19	36	30
Iceland	..	..	..	–99	–95	–99*	0.6	3	0.5*
Norway	..	..	..	–58	–56	–60	28	38	32
Portugal	..	..	..	–49	–34	–41*	46	59	49*
Spain	..	..	..	–67	–44	–46*	37	43	34*
Sweden	89	97	95	–14	–5	–4	73	78	67
Switzerland	..	..	..	28	24	22	185	160	128
Turkey	..	..	..	–96	–91	–77	11	9	10

* 1983.
1. Excluding inter-EEC trade.

$$SN = 100 \times \frac{Output}{Domestic\ demande} = 100 \times \frac{Output}{Output - X + M}$$

$$IS = 100\ \frac{X - M}{X + M} = 100\ \frac{C - 1}{C + 1}$$

X : Exports
M : Imports
C : X/M
X_{im} : Manufacturing exports

$$CA_{ij} = 100 \times \frac{X_{ij}}{\sum_i X_{ii}} \bigg/ \frac{X_{im}}{\sum_i X_{im}}$$

Of the eleven countries for which full data are available, only the five countries with a positive balance and the EEC had a "niche" specialisation indicator consistently above 100. In 1980 specialisation seemed to be greater in Japan and Germany than in the United States. It was on the increase in many countries and in the EEC between 1970 and 1975. But this uptrend persisted only in Japan after 1975, in contrast to all other countries.

The intra-industry specialisation ratio for the period as a whole is positive for the same countries, the EEC, Switzerland and, from 1981, Ireland. The most important trends concern:

– Japan, the country showing the largest relative surplus and steepest uptrend over time;
– The decline in intra-industry specialisation in the other major countries, except France where the ratio remained quite low. It is also worthwhile noting that for the first time both the trade balance for highly R&D intensive industries and intra-industry specialisation ratio turn negative in the United States from 1984 and in the United Kingdom from 1983.

If these performances are converted into export/import ratios, the following were the front runners in the trade of high intensity R&D industries in 1982 and 1984:

	Export/import ratio	
	1982	1984
Japan	4.7 or 470%	5.6 or 560%
United States	1.4 or 140%	0.9 or 90%
EEC	1.1 or 110%	1.0 or 100%

These results reflect the strength of the Japanese position and the decline in United States competitiveness due partly to the major strenthening of the dollar.

c) *Apparent (or revealed) comparative advantage (CA)*

In assessing a country's specialisation, its exports from the industries under review must be compared with total exports in order to take into account the effects of scale. This is why an indicator of apparent comparative advantage (CA) – also known as revealed comparative advantage – has been established.

It is the share of an industry in a country's total exports of manufacturing industry, compared with the same indicator for all OECD countries, which equals 100[20].

In consequence, if the coefficient obtained exceeds 100, it indicates a comparative advantage and, if it is less than 100, it means a disadvantage and weak specialisation (see Table 2.16).

Only three of the major countries have a comparative advantage in high-intensity R&D industries:

– The United States has the highest degree of specialisation;
– Japan's marked uptrend after 1975 has given it second place;
– The United Kingdom ranks third, mainly due to the improvement in its advantage between 1970 and 1975.

As in the case of trade balances and inter-industry specialisation in the three large countries, Switzerland and Ireland should be added. However, while the comparative advantage of Switzerland has declined steadily since 1974, that of Ireland continues to grow thanks to its exports of electronic equipment and computers originating from foreign firms, for the most part American, which have been set up there. The ratio of these exports to manufacturing exports has seen a spectacular growth in recent years. Nevertheless, the improvement of the comparative advantage of Ireland is not allied to an exceptional technological effort on its part in that the technology required for its electronic exports has been imported.

No other country, including Germany (declining advantage) and France, has a comparative advantage in high R&D intensity industries.

The range of indicators and variables used to analyse the position of countries in international trade involving high R&D intensity industries lead to four main conclusions:

– Despite a decline in intra-industry specialisation due to a large increase in imports, the United States still leads the field in the dissemination of technology through the export of goods;
– Japan ranks second in the world thanks to a sustained uptrend;
– Divergent trends are found in the major European countries;
– Overall, this results in a decline for the EEC as regards high intensity R&D industries particularly after 1980.

R&D and trends in industrial structures

It is not possible, in such a short report, to analyse in detail how the high R&D intensity industries serve as pace setters for industry as a whole due to their own drive and to the indirect effects resulting from interdependence between industries. We shall limit our comments to two areas in the relationships between R&D and growth and between R&D and productivity and employment.

The relationship between R&D and growth

The R&D-growth relationship can be considered at individual industry level or at overall level for each country.

i) An analysis of the trend in all industries in the various countries does confirm the assumption that high-R&D intensity industries are growth industries. This is because they are in the first phase of the innovation cycle and are able to keep themselves there by introducing successive product innovations, each of which has a quite short product life cycle.

To calculate the growth rate of each industry at the OECD level, deflators specific to each industry and to each country have been used. Thus, Table 2.17 shows that the industries with the highest growth do come from the high-intensity category (electronics, computers, drugs and scientific instruments). Other industries, in the medium-intensity category, also have relatively high growth rates (automobiles, chemicals). The low-growth industries are all low R&D intensity ones. Considerable differences in growth, thus, exist between industry categories.

Table 1.17
Average annual growth of output (volume)
1970-1980

High growth		Medium growth		Low growth	
Electronics	8.1	Electrical machinery	2.8	Oil refining	1.7
Computers	7.5	Food	2.8	Wood/cork/furniture	1.6
Drugs	6.8	Shipbuilding	2.4	Ferrous metals	1.4
Automobiles	5.7	Other manuf. industry	2.3	Non-ferrous metals	1.3
Chimicals	5.0	Paper-printing	2.1	Manufacturing of metal	1.3
Instruments	4.9	Stone, clay, glass	2.0	Textiles, footwear/leather	0.8
Rubber/plastics	4.3				

ii) If we compare growth in industrial output with growth or R&D expenditure, we find two different situations (see Graph 3 in Annex).

In the first group of countries – Italy, Canada, the United States and the Netherlands – output grew at roughly the same rate as R&D expenditures.

In the second group (Belgium, Sweden, Germany and, possibly, France) R&D grew more rapidly than output.

Two extreme cases can be defined:
– Japan which has the highest growth rates in both output and R&D;
– The United Kingdom where low R&D growth is accompanied by a decline in industrial output.

It should be quite obvious that the overall link between R&D and industrial growth in each country closely depends on the structure of industrial output: the greater the share of rapid growth industries, the greater overall growth will be. This is no doubt the case of the countries generating growth in electronics (Japan and Sweden) and computers (United States, Japan, Italy and the Netherlands). It should, however, be noted that these industries are as yet responsible for only a very low share of industrial production and that their main contribution to growth comes via the modernisation of production in other industries.

The relationship between R&D productivity and employment

By and large, apparent labour productivity increased in the 1970-1980 decade, especially in Japan (+6.9 per cent a year on average), Belgium (+5 per cent), Italy (+3.8 per cent) and the Netherlands (+3.6 per cent). However, when both employment and productivity are taken into consideration, there are significant differences between countries (Table 2.18). Moreover, the role played by R&D intensive industries in these trends is not absolutely clear. Furthermore, there is no one explanation for the more general impact of R&D on trends in productivity. These three points are examined in turn.

i) A comparison between trends in employment and in productivity reveals three groups of countries:
– A first group in which productivity growth is due to a high growth in output combined with a less rapid decline in employment (Japan, United States, Canada and Italy);
– A second group in which productivity growth is on the contrary due to a major fall in employment despite a modest decrease or slight increase in output (United Kingdom, Belgium);
– A third group in which the productivity trend is explained by the fact that output is increasing and employment decreasing at similar rates (Germany, France, Australia, the Netherlands and Sweden).

These results suggest that there is a higher degree of substitution of capital for labour in the second group than in the two others.

ii) The effect of the structure of industrial output on labour productivity is neither clear nor direct.

Table 2.18
Growth of manufacturing industry
Annual average growth 1972-1981

	Output	Apparent labour productivity	Number of workers employed	Number of researchers
United States	2.6	1.3	+0.6	+4.7
Japan	4.1	6.9	−0.4	+5.1
Germany	1.2	3.0	−0.9	+3.4
France	1.2	3.4	−1.6	+2.7
United Kingdom	−1.3	1.6	−1.9	−0.2
Italy	3.2	3.8	−0.8	+3.9
Canada	2.8	0.8	+1.1	+4.5
Australia	1.1	2.4	−1.2	..
Netherlands	2.1	3.6	−1.7	+3.4
Sweden	0.7	1.3	−0.5	+6.8
Belgium	1.5	5.0	−3.5	+2.9

First, the proportion of the labour force working in high R&D intensity industries has risen, while that in low-intensity industries has fallen (Table 34 in Annex).

Secondly, labour productivity seems to be higher in the mature industries than in high R&D intensity ones. As the weight of the latter in the labour force rises, the effect on average productivity is therefore negative. However, as productivity grows more rapidly in high-intensity industries, there is also a beneficial effect in dynamic terms.

iii) A comparison of trends in productivity and in R&D (as measured by the number of researchers) shows that situations differ (Table 2.18 and Graph 4 in Annex).

Japan combines the sharpest increase in productivity with quite high R&D growth. But the United States, Canada and Sweden, where R&D growth was equally high or higher, barely managed a third of Japan's productivity growth.

A number of countries combine similar growth rates for productivity and number of researchers (France, Belgium, Italy, Germany and the Netherlands). The United Kingdom is noteworthy in that a slight decrease is recorded in the number of researchers.

Once again, Japan's dynamism stands in contrast to that of the other advanced countries. But it must be stressed that this dynamism is not limited to R&D variables; so far, we have been speaking about industrial growth and labour productivity. It may well be that factors which were presented in Part I as evidence of Japanese technological achievements, are to be partly explained by more general factors relating to competitiveness.

If we examine the actual level of labour productivity and research efforts (rather than their growth over the 1970-80 period), we find that the United States still led the field in 1981 despite the tremendous effort by Japan and some other European countries during the 1970s to reduce this lead (see Graph 5 in Annex).

R&D AND TRENDS IN COMPETITIVENESS

Examining the link between the production of technology (as measured by R&D) and industrial competitiveness, enables us to go further than in Part I in analysing the results obtained by countries in the dissemination of technology. As indicated earlier, the link between technological performance and industrial competitiveness works both ways. In analysing this interdependence, it is necessary to take into account the effect of all the factors explaining competitiveness in all industrial activities, whatever the degree of technological intensity.

Differing Trends

Before examining these factors and their effects, it is worthwhile stating the purpose of this section of the report which is to identify and present countries' trade situations for all categories of industry, and not just for the high R&D intensity group.

The trade balances by category of industry (see Graph 6 in Annex) suggest that the trends differ quite considerably from one country to another:

– Japan is the only country to show a surplus for all three categories of industry, with absolute values twice and three times those of the United States in the high-intensity and medium-intensity industries respectively. These results are due to a tremendous increase in the balances during the reference decade.

– Only five countries show a surplus for high-intensity industries combined with a deficit for low-intensity industries: the United States, Germany, France, the United Kingdom and Switzerland.

– The United States is the only country whose trade balance in high-intensity industries is higher than in medium-intensity industries. In other words, the medium intensity industries play a major role in keeping trade in balance in many of the countries studied.

Tableau 2.19
Recapitulation: Trade balance of manufacturing industries according to their R&D intensity 1970-1984

	High intensity	Medium intensity	Low intensity		High intensity	Medium intensity	Low intensity
United States	positive[1]	positive[2]	negative	Canada	negative	negative[3]	positive
Japan	positive	positive	positive	Australia	negative	negative	positive[4]
EEC*	positive	positive	positive[14]	New Zealand	negative	negative	positive
Germany	positive	positive	negative	Austria	negative	negative	positive
France	positive[6]	positive	negative[7]	Finland	negative	negative	positive
United Kingdom	positive[13]	positive[13]	negative	Iceland	negative	negative	negative
Italy	negative[10]	positive	positive	Norway	negative	negative	negative
Netherlands	negative[11]	positive	positive	Portugal	negative	negative	positive
Belgium-Luxembourg	negative	positive	positive	Spain	negative	positive[12]	positive
Denmark	negative	negative	negative[5]	Sweden	negative	positive	positive
Ireland	positive[8]	negative	negative[9]	Switzerland	positive	positive	negative
Greece	negative	negative	negative	Turkey	negative	negative	negative

* Excluding inter-EEC trade.
1. Negative in 1984.
2. Negative from 1982.
3. Positive from 1981.
4. Negative in 1981, 1982 et 1984.
5. Negative to 1980.
6. Negative in 1970, 1972, 1973 and 1974. If one takes into account exports which are distributed abroad by country of origin the balance is negative over the period with the exception of the end of the 1970s.
7. Positive before 1980.
8. Negative to 1980.
9. Positive to 1978.
10. Except 1970-72, 1975-78 and 1983 where it is positive.
11. Except 1976 and 1977 where it is positive.
12. Negative to 1980.
13. Negative in 1983 and 1984.
14. Negative in 1979, 1980, 1983 and 1984.

— Most of the countries with deficits in high-intensity industries have positive balances in low-intensity industries (Italy, Canada, Australia, the Netherlands, Sweden, Belgium, Denmark, New Zealand, Austria, Finland, Portugal and Spain). Only five of these countries (Italy, the Netherlands, Belgium, Spain and Sweden) have positive balances for medium-intensity industries. For all the rest, the low-intensity category is the only one where balances are positive; but these are generally higher than in medium-intensity industries.
— The trend for the EEC as a whole shows highly specific characteristics: only the medium-intensity industries have an increasing positive balance over time; the low-intensity industries show a downtrend, while the surplus of the high R&D intensity industries has grown only very slightly.
— Finally, some small countries (Greece, Iceland, Norway and Turkey) have negative balances in all three categories of industry (see Table 2.19 and Graph 6 in the Annex).

Table 2.19 recapitulates these trade positions for the 1970-1984 period as a whole.

Factors in competitiveness

The trends in trade identified can be explained not only by research efforts or by the effectiveness of the production of technology but also by a whole series of other common factors related to competitiveness. Some of these factors of a more qualitative kind (such as distribution networks, after-sales services, schemes for building client loyalty, delivery terms, etc.) are difficult to measure and compare. Others, such as the trend in costs and prices and in demand elasticities are easier to quantify.

Trends in unit labour costs and relative export prices

Since the start of the 1980s, the rise in wages has slowed down in the OECD countries, even in those where growth was generally above average (see Table 35 in Annex).

The trend in comparative unit costs in common currency terms widened the cost range between the five major OECD countries between 1970 and 1979 (see Graph 7 in Annex). Since then, the relative downturn in labour costs in Japan, together with the rise in the United Kingdom and the United States, has brought the growth curves closer. It should be noted that the growth in labour costs has been very moderate everywhere since 1982.

In assessing the impact of labour costs on a country's trade ranking, it is not enough to consider the trend in these costs as compared with the costs of competitors. Two other kinds of information are required: the comparative level of labour costs and the impact of these costs on export prices, which depend, among other things, on the structure of exports by commodity.

Trends in relative export prices supplement the incomplete picture given by trends in labour costs.

Export prices are determined by both domestic factors, (i.e. trends in wholesale prices and, therefore, in costs) and international factors, more specifically the trend in competitors' export prices. These two sets of factors do not work independently of each other.

The relationship between the two depends on how mark-ups are fixed by exporters. The comparative trends in domestic wholesale prices and in competitors' export prices give the exporter differing degrees of latitude in choosing between a change in his mark-up and a change in his market share. In a country where prices are low, exporters are freer to pass on cost increases or raise their mark-ups, depending on whether the cost trend is unfavourable or favourable. In the opposite case, where the prices are high and costs are rising more rapidly than those of competitors, the exporter may either reduce his mark-up to maintain his relative price or else lose market shares.

The trend in the relative export prices of manufacturing industries (Table 36 and Graph 8 in Annex) reflects trends in both costs and mark-ups. As previously, the years 1979-80 seem to mark a turning point, with the trends in the relative prices of major countries tending to draw closer. The changes in country positions seem to suggest that, at least in terms of the trend[21], the effect of prices was not decisive throughout the reference period.

A comparison between relative labour costs and relative export prices shows that changes in the latter were more limited. This is not surprising for the following reasons.

First, labour accounts for only a part of the cost price and, secondly, the price set for exports includes the mark-up which itself depends on the trend in competitors' prices. Allowing for these disparities, trends in relative export prices up to 1983 were more favourable than those in relative labour costs in seven countries out of ten: Japan, Germany, France, the United Kingdom, Italy, Canada and the Netherlands.

In the United States, relative prices followed a distinctive pattern. Because of changes in exchange rates between 1980 and 1982, relative export prices suddenly rose at a time when those of the other major economies were declining, a clear pointer to a downturn in the competitiveness of United States' products.

Demand elasticities

The elasticity of exports with respect to foreign demand and of imports with respect to domestic demand have been calculated for the 1970-1980 period for each category of industry (high, medium and low R&D intensity) and for the eleven EEC countries. The results are given in Graphs 9, 10 and 11 in the Annex.

Elasticities are designed to measure the degree to which exports (imports) vary in response to a change in foreign demand (domestic demand). On the export side, they measure exporters' capacity to respond (to adjust to and even anticipate) an increase in demand on the world market. The higher the elasticity, the greater is this capacity.

Conversely, on the import side, elasticities with respect to domestic demand are a measure of the way in which foreign suppliers adjust to the trend in domestic demand. It, therefore, enables us to ascertain the extent to which the domestic market is exposed to import penetration.

It must be noted here that the significance of demand elasticities differs not only as between imports and exports but also in relation to the dynamism of the markets under consideration. In the case of exports, therefore, it is essential to have a high export/demand elasticity for the high R&D intensity industries, as growth in international demand will lead to growth in exports.

On the import side, however, it is better to have a low demand elasticity in this same category – so as to limit import growth – together with a high elasticity in declining industries. A combination of imports in declining activities with exports in buoyant industries is a structural advantage as regards the trade balance which is, thus, tilted towards a surplus.

As shown by the recapitulative Table 2.20 (for further details see Tables 37 and 38 in Annex), import/demand elasticities were generally higher than those for exports in the 1970-1980 period. Japan stands out with the highest export/demand elasticity and one of the lowest import/demand elasticities.

But the most instructive comparison of the two elasticities is by category of industry (see Graphs 9, 10 and 11 in Annex).

In the high R&D intensity industries, Japan is the best placed with an export/demand elasticity higher

Table 2.20

Elasticities of exports with respect to foreign demand and elasticities of imports with respect to domestic demand (Total manufacturing industries) 1970-1980

	Elasticities of exports	Elasticities of imports
United States	1.01	1.35
Japan	1.09	1.15
Germany	1.01	1.36
France	1.07	1.24
United Kingdom	0.94	1.49
Italy	1.05	
Canada	0.74	1.14
Australia	0.81	1.20
Netherlands	0.99	1.13
Sweden	0.89	1.19
Belgium	0.97	1.22
EEC	1.01	1.36

than that for imports. The United States is in the opposite position, with a higher import elasticity. In the other countries, with the exception of Canada, the export elasticity is close to 1, which implies an export growth parallel to market growth (constancy of market shares). In all countries except Japan (and to a lesser extent France, Australia and the Netherlands), import elasticity is higher than that for exports, which means the domestic market penetration is speeding up.

In the category of medium-intensity industries, Japan is once again the most competitive country. With similar export and import elasticities greater than 1, France and the Netherlands show increases in both export market shares and import penetration. In all the other countries, especially in the United Kingdom, Italy and the United States, the uptrend is sharper for import penetration than for export shares.

The category of low-intensity industries in all countries has demand elasticity values that are higher for imports than for exports. Italy is the only country whose exports show some dynamism with elasticity over 1. Germany, France, the Netherlands and the EEC have maintained or increased their export shares, with an export/demand elasticity of close to 1.

Trends in industrial competitiveness

An economy's industrial competitiveness can be measured by two indicators: the export market share and the rate of import penetration on the domestic market. These two variables relate to the capacity of products to stand up to foreign competition either on the international or domestic market. If it is to be at all meaningful, the analysis should be of a dynamic type, with the trend in the indicators over time providing information on the competitiveness of products and enterprises in each economy.

Competitiveness is quite often confused to some extent with specialisation. Without going into a semantic and theoretical discussion, it may be said that specialisation provides a measure of the propensity of one country more than the others to export particular products.

Competitiveness is a more dynamic measure of the capacity of products to win a share of the export market: it may concern products as a whole, in which case specialisation is not affected, or only some. In the more likely second case, the kind or kinds of specialisation will be modified by differential competitiveness. The two concepts are, therefore, obviously related.

Before describing the competitiveness indicators, it may be worthwhile reviewing specialisation by countries in the various industry categories.

Geographical breakdown of output and specialisation

a) *Output*

The breakdown of industrial output among the various countries changes during the decade 1970-1980 (see Tables 39 and 40 in Annex). The main features of the changes varied according to industry category and were as follows:

i) A downtrend in the United States' share in all R&D intensity categories, especially in the high-intensity industries where it was most marked, a downtrend that must be linked to changes in the relative weights of OECD economies. Electrical machinery, electronics and aerospace industries were particularly affected in the high R&D intensity category.

ii) The general growth in Japan was particularly marked in the high-intensity industries (especially computers and instruments) and also in the low-intensity industries.

iii) In Europe, France and Germany showed growth in high-intensity industries (thanks to aerospace) while the United Kingdom's shares were unchanged (high-intensity) or down (medium and low-intensity). In the EEC as a whole, output is up in the high-intensity industries (and also to a lesser extent in medium-intensity industries), while it remained at the same level in low-intensity industries.

The changes in the geographical breakdown of industrial output concern the international location of the various industries and, accordingly, simply indicate potential changes in international trade specialisation.

b) *Specialisation*

Various indicators can be used to analyse specialisation in international trade, but the most significant are the apparent (or revealed) *comparative advantage* (CA) which has already been used (see section above) and *specialisation by "niche"* (SN). Apparent comparative advantage[19] enables us to calculate whether a country exports proportionally more of one good than others, account being taken of the average structure of trade by goods (for example, the weight of the good in exports), or whether a country exports proportionally more of a particular good than other countries, account being taken of the geographical structure of exports (for example, the weight of this country in total exports). The comparative advantage indicator, therefore, points to the main trend in exports.

The indicator of specialisation by "niche" measures the degree of a country's independence given the size of its internal demand. A country is increasingly specialised in an industry (or product) to the degree that it produces more than it consumes.

Tables 2.21 and 2.22 give the values for the comparative advantages and the specialisation by area of the various countries in the three industry categories for the 1970-1984 and 1970-1980 period respectively and two observations may be made as regards:

i) Specialisation in industry categories:

Only five countries had a comparative advantage in the high R&D intensity industries: the United

Table 2.21
Apparent comparative advantage of manufacturing industry
OECD average = 100

	High R&D			Medium R&D			Low R&D		
	1970	1980	1984	1970	1980	1984	1970	1980	1984
United States	158	156	156	109	106	98	63	64	64
Japan	123	141	147	78	105	101	114	75	68
EEC[1]	93	93	82	105	101	99	97	102	114
Germany	96	93	82	124	117	119	76	82	86
France	85	81	83	94	98	95	110	108	114
United Kingdom	104	121	118	117	108	98	81	80	89
Italy	77	62	56	99	91	90	111	128	141
Netherlands	97	76	62	62	71	74	139	143	156
Belgium-Luxembourg	44	47	37	94	98	101	127	125	137
Denmark	72	75	66	62	63	60	150	154	171
Ireland	71	113	159	21	56	54	192	146	120
Greece	14	17	13	60	35	31	176	211	244
Canada	55	48	43	124	108	127	91	112	100
Australia	17	26	19	66	71	78	160	166	169
New Zealand	4	9	9*	9	24	29*	233	231	234*
Austria	69	73	63	72	79	85	141	137	142
Finland	19	36	30	36	50	48	199	192	210
Iceland	0.6	3	0.5*	64	46	76*	176	208	183*
Norway	28	38	32	90	93	100	139	136	142
Portugal	46	59	49*	36	34	40*	188	196	199*
Spain	37	43	34*	62	86	83*	165	143	157*
Sweden	73	78	67	83	87	88	128	125	136
Switzerland	185	160	128	105	113	114	61	57	64
Turkey	11	9	10	46	35	28	192	218	249

* 1983.
1. Excluding inter-EEC trade.

Table 2.22
Specialisation by "niche" of manufacturing industries

	High intensity			Medium intensity			Low intensity		
	1970	1975	1980	1970	1975	1980	1970	1975	1980
United States	105	110	109	101	105	102	97	97	
Japan	109	114	126	104	112	118	106	106	102
Germany	118	121	115	127	145	137	98	102	95
France	100	104	102	101	109	104	101	102	100
United Kingdom	108	113	108	112	114	109	96	94	96
Italy	99	99	94	106	117	105	102	107	106
Canada	81	73	66	103	83	90	104	101	108
Australia	61	66	64	82	83	84	103	103	102
Netherlands	93	100	97	78	100	105	105	114	108
Sweden	89	97	95	97	100	107	104	102	103
Belgium	77	81	79	102	104	98	122	116	114
EEC	106	110	104	112	121	115	100	101	99

States, Japan, the United Kingdom, Switzerland and Ireland. These countries, with the exception of Ireland, have slight comparative advantages in the medium and low-intensity industries.
Germany has the most clear-cut advantage in the category of medium-intensity industries thanks to its strong position in machinery, motor vehicles and chemicals.
All the other countries, including France, have a comparative advantage in the low-intensity industries but, other than in the case of Canada for medium-intensity, this is their only advantage.

On the other hand, in the case of the specialisation compared with internal demand (specialisation by "niche") in only five countries, of the eleven available, do high-technology industries produce more than internal demand, with Japan in the lead followed by Germany. These are also in the lead for medium-intensity industries with Germany preceding Japan.

ii) Trend in specialisation

In the high-intensity industries, three countries (Japan, the United Kingdom and Ireland) have improved their comparative advantage whereas Switzerland has seen its decrease. Throughout the period, the United States has maintained its comparative advantage which is the highest of the OECD area. All indicators confirm that the United States is highly specialised in industries of this group.

There are two diametrically opposed trends in the medium and low-intensity industries in Japan which has acquired comparative advantage, noteworthy (while certainly limited), in that there was no initial specialisation. This is clearly a reversal of the trend since the initially specialised low-intensity industries have taken the opposite track.

Unlike Japan, the United States, and the United Kingdom in particular, have lost their comparative advantage in medium-intensity industries. Specialisation in Germany has also diminished slightly.

Japan was the only country where area specialisation of high-intensity industries grew, reflecting the major increase in its trade surplus. Growth in area specialisation of the medium intensity industries was most marked in the Netherlands and Germany followed by Japan and Sweden.

In sum, the most marked changes in specialisation would seem to be:
- an increase in Japan for high-intensity industries, and
- an increase in Germany, the Netherlands and Sweden in medium-intensity industries.

These changes in international specialisation relate, among other things, to differing trends in competitiveness.

Measurement of industrial competitiveness

Three indicators will be used: export market shares, import penetration rates and rates of exposure to foreign competition. These provide information on both exports and imports.

Table 2.23

Export market shares of manufacturing industry

With intra-EEC flows adjusted

	High R&D			Medium R&D			Low R&D		
	1970	1980	1984	1970	1980	1984	1970	1980	1984
United States	35.4	30.5	31.2	26.0	22.5	20.5	16.1	15.0	14.3
Japan	15.0	21.3	28.8	10.1	17.1	21.5	15.7	13.7	15.5
EEC	33.0	33.4	26.1	40.1	39.4	33.9	34.4	37.9	34.8
Germany	10.7	10.5	8.0	15.9	14.7	13.5	8.5	9.9	8.9
France	4.7	5.8	5.3	5.0	6.3	5.3	5.6	7.2	6.3
United Kingdom	8.2	8.8	6.0	10.2	8.5	5.4	7.7	5.8	4.7
Italy	3.1	2.9	2.8	4.6	5.1	4.9	4.9	6.5	6.8
Netherlands	4.2	3.1	1.9	1.3	1.6	1.7	2.9	3.5	2.9
Belgium-Luxembourg	0.8	0.9	0.6	1.9	2.1	1.9	2.3	2.3	2.0
Denmark	0.8	0.7	0.7	0.8	0.7	0.7	1.8	1.4	1.8
Ireland	0.1	0.2	0.5	0.03	0.1	0.2	0.2	0.3	0.4
Greece	0.01	0.0	0.03	0.04	0.1	0.08	0.2	0.6	0.5
Canada	4.4	2.5	3.3	10.7	6.2	10.2	8.4	7.1	8.6
Australia	0.3	0.3	0.2	1.2	1.0	1.0	3.2	2.8	2.4
New Zealand	0.03	0.05	0.05	0.06	0.1	0.1	1.8	1.5	1.5
Austria	1.2	1.4	1.1	1.3	1.7	1.6	2.7	3.3	2.9
Finland	0.2	0.5	0.4	0.5	0.8	0.8	3.2	3.7	3.6
Iceland	E	E	E	0.01	0.03	0.03	0.08	0.1	0.1
Norway	0.4	0.4	0.3	1.3	1.1	1.0	2.2	1.8	1.6
Portugal	0.3	0.3	0.2	0.2	0.1	0.2	1.1	1.2	1.2
Spain	0.5	0.9	0.7	0.8	2.0	1.8	2.3	3.7	4.0
Sweden	3.0	2.7	2.2	3.6	3.3	3.1	5.9	5.2	5.1
Switzerland	5.9	5.3	3.8	3.6	4.0	3.5	2.2	2.2	2.1
Turkey	E	E	E	0.02	0.04	0.1	0.1	0.3	1.7
Total OECD	100	100	100	100	100	100	100	100	100

* Calculations based on current dollars.

Table 2.24

Export market shares of manufacturing industry

Without adjustment for intra-EEC flows

	High R&D			Medium R&D			Low R&D		
	1970	1980	1984	1970	1980	1984	1970	1980	1984
United States	28.3	24.1	25.6	19.5	16.4	15.8	11.3	9.8	9.9
Japan	12.0	16.8	24.5	7.6	12.5	16.5	11.1	9.0	10.7
Germany	16.0	15.6	12.9	20.7	19.7	18.5	12.6	13.8	12.8
France	7.0	7.7	6.9	7.7	9.3	7.8	9.0	10.3	9.0
United Kingdom	9.5	10.8	8.3	10.7	9.6	6.7	7.6	6.3	5.9
Italy	5.0	4.4	3.9	6.4	6.4	6.2	7.1	9.1	9.4
Netherlands	5.0	4.3	3.3	3.2	4.0	3.9	7.1	8.1	7.8
Belgium-Luxembourg	2.4	2.5	1.7	5.2	5.2	4.4	7.0	6.6	5.7
Denmark	1.0	1.0	0.9	0.9	0.8	0.8	2.2	2.0	2.2
Ireland	0.2	0.7	1.3	0.09	0.3	0.4	0.6	1.0	0.9
Greece	0.02	0.06	0.05	0.1	0.1	0.1	0.3	0.7	0.7
Canada	3.5	2.0	2.7	8.0	4.5	7.9	5.9	4.6	5.9
Australia	0.2	0.2	0.2	0.9	0.7	0.7	2.2	1.8	1.7
New Zealand	0.02	0.03	0.04	0.05	0.09	0.12	1.2	1.0	1.0
Austria	0.9	1.1	0.9	1.0	1.2	1.2	1.9	2.1	2.0
Finland	0.2	0.4	0.4	0.3	0.4	0.6	2.2	2.4	2.5
Iceland	0.0	0.0	0.0	0.0	0.02	0.03	0.05	0.1	0.07
Norway	0.2	0.3	0.3	1.0	0.8	0.8	1.5	1.1	1.1
Portugal	0.1	0.1	0.2	0.1	0.1	0.1	0.8	0.6	0.7
Spain	0.2	0.6	0.7	0.6	1.5	1.4	1.6	2.4	2.7
Sweden	2.4	2.1	1.8	2.7	2.4	2.4	4.1	3.4	3.5
Switzerland	4.7	4.2	3.1	2.7	2.9	2.7	1.5	1.5	1.4
Turkey	0.0	0.0	0.0	0.02	0.01	0.1	0.1	0.2	1.2
Total OECD	100	100	100	100	100	100	100	100	100

* Calculations based on current dollars.

a) Export market shares

These measure the geographical breakdown of exports for each category of industry (see Tables 2.23 and 2.24). The main point to be noted is the diminishing share of export markets of the United States and the EEC and the increasing share of Japan.

The *United States'* market has declined in all three categories, especially in medium-intensity industries. However, in spite of the strength of the dollar between 1983 and 1984, exports of high-intensity technology have grown faster than the OECD average. The United States is now the major world exporter in such industries (see Table 2.24).

Japan's export market shares have doubled in the high and medium-intensity industries but have fallen slightly in low-intensity industries. This reflects Japanese industrial specialisation in high technology industries and strong growth rates.

The *EEC*'s export market shares have fallen for high and medium-intensity industries while remaining stable for low-intensity industries. The losses in the first category could be attributed, for the most part, to Germany, the United Kingdom and the Netherlands and those in the second category to the United Kingdom and, to a lesser extent, Germany. Although in overall deficit, low-intensity industries have maintained their market share; however, the reduced specialisation of EEC countries in high-intensity industries may, in time, create difficulties given that trade liberalisation in this area will lead to keener competition with Third World countries and growth prospects for these industries are modest.

b) The import penetration rate

This rate measures the share of imports in domestic demand, which in turn reflects the degree of exposure of the domestic market and the competitiveness of national products. It is, therefore, necessary to take account of the degree of exposure of each market, which is particularly marked in small European countries such as Belgium and the Netherlands, but relatively limited in the United States and Japan (see Table 2.25).

The general rise in penetration rates is evidence of increased overall internationalisation of trade. However, this trend appears more pronounced in the high R&D intensity industries (which reflects the tendency mentioned earlier whereby a higher proportion of output is exported). Each country's position and the relevant changes have to be assessed in the light of this trend.

Analysis of these penetration rates by and large confirms the conclusions already drawn for the export

Table 2.25
Rate of import penetration of manufactures

	High R&D			Medium R&D			Low R&D		
	1970	1975	1980	1970	1975	1980	1970	1975	1980
United States	4.8	7.7	14.3	7.0	8.9	11.2	4.8	5.7	6.1
Japan	6.6	6.4	7.9	5.5	5.0	6.9	3.5	4.4	5.4
Germany	21.6	28.4	42.5	22.8	27.4	33.4	16.8	21.6	27.4
France	23.1	23.3	28.8	23.7	26.0	31.6	11.8	14.7	17.7
United Kingdom	18.6	31.2	44.2	22.0	29.3	43.1	13.4	17.4	18.4
Italy	18.3	25.3	33.4	23.8	27.7	41.8	11.9	17.8	25.4
Canada	42.6	45.7	58.0	58.5	59.2	59.2	12.7	14.1	13.8
Australia	42.2	39.7	43.1	30.6	30.4	32.5	14.2	14.8	18.5
Netherlands	71.0	67.8	69.8	80.5	79.3	85.6	38.3	43.4	51.0
Sweden	44.9	46.1	54.6	44.4	48.1	53.5	23.1	27.2	28.3
Belgium	75.7	87.3	..	88.0	86.3	..	40.3	48.8	64.0
EEC	25.1	32.1	41.7	28.1	33.3	42.4	16.0	20.5	25.4

market shares (Graph 12 in the Annex uses both indicators for the high-intensity category of industries). Thus:

- the competitiveness of United States' products has diminished on both the domestic and international markets;
- only in Japan, and to a lesser extent France, have all indicators of competitiveness improved at the same time in the high R&D intensity industries.

c) *Degree of exposure to foreign competition*

This third indicator measures, as it were, the degree of internationalisation of trade specific to each category of industry in the various countries[22] (Table 2.26).

As regards categories of industry, the degree of internationalisation increased the most in the high-intensity group between 1970 and 1980. At the end of the period, exposure to competition seemed much greater in the high and medium-intensity industries than in the low-intensity industries.

As regards countries or areas, the EEC countries individually or as a whole, seem to be far more exposed to international competition than Japan or the United States which are on comparable levels. The Netherlands is an extreme case since it has the highest degree of exposure in each industry category. (These rates reflect not only the small size of the national market but also the fact that the Netherlands is an important transit centre for trade.) Conversely, (outside the United States and Japan) France and Australia show relatively moderate degrees of exposure to foreign competition in high or medium-intensity industries.

Combined analysis of all these indicators of competitiveness brings out two major points:

- The general trend in relative competitiveness has been particularly favourable in Japan in all industries. Conversely, the competitiveness of United States' and EEC products has diminished, although the consequences differ: the EEC's positions seem to be in greater danger owing to greater exposure and less advantageous trade specialisation (i.e. one geared less to expanding industries);
- The trends are more pronounced in high R&D intensity industries than in the other industries owing to their greater internationalisation, as evidenced by generally high rates for exports and exposure to competition. Losses or gains in competitiveness seem to be amplified by the intensity of trade.

Table 2.26
Degree of manufacturing industry exposure to foreign competition

	High R&D			Medium R&D			Low R&D		
	1970	1975	1980	1970	1975	1980	1970	1975	1980
United States	14.6	22.8	32.9	15.1	21.1	23.1	7.5	9.1	9.7
Japan	19.5	23.4	32.8	14.9	19.9	27.2	11.8	14.0	12.8
Germany	48.1	57.8	71.4	53.3	63.7	67.7	29.4	38.7	44.8
France	41.1	43.8	50.4	42.8	50.0	55.2	23.0	26.9	32.4
United Kingdom	38.8	58.5	71.2	46.2	56.5	70.5	22.3	28.0	30.7
Italy	33.2	43.9	52.9	45.4	55.5	67.9	24.4	37.3	47.8
Canada	59.8	59.9	73.4	83.3	80.0	81.5	27.0	27.2	31.2
Australia	45.8	45.2	49.7	41.9	41.8	45.8	29.0	30.2	35.4
Netherlands	91.0	89.7	90.6	95.1	95.8	98.0	63.9	72.0	77.0
Sweden	66.1	70.2	78.4	68.1	73.3	79.9	43.2	52.7	50.1
EEC	47.1	57.8	67.6	53.9	63.3	71.1	29.8	38.0	44.3

NOTES AND REFERENCES

1. Throughout the report, comparisons are made between countries using implicit GDP purchasing power parities provided by the OECD National Accounts Division. For further details, see *OECD Science and Technology Indicators Resources Devoted to R&D*, Paris, 1984 – General Technical Notes, points 2 and 3.

2. OECD holds major R&D surveys only every two years and 1981 is the latest year for which a full set of data is available at the time of writing. However, international data for 1983 have, insofar as possible, been used at the level of national and sector totals and, for some countries, 1984 and even 1985 data have also been used. The data for GERD and its components will be found in "Science and Technology Indicators Basic Statistical Series – Volume B, Gross National Expenditure on R&D GERD 1969-1982" (Paris, March 1985). Other selected, more recent or varied indicators are issued in the report "Recent Results". The present report concentrates on the period since 1979 as the decade 1969-1979 was dealt with at length in *OECD Science and Technology Indicators Resources Devoted to R&D* (Paris, 1984).

3. Throughout this report, trends in R&D spending are measured in volume, i.e. the data have been adjusted to exclude the effects of inflation. The deflator used is the implicit GDP price index (base 1975 = 100) as published by the OECD National Accounts Division and updated from the latest edition of the OECD Economic Outlook.

4. A special characteristic of R&D indicators for Japan is that, as far as the personnel series are concerned, data are expressed in number of persons *"regularly"* employed in the research category and in number of persons *("head-count")* in the other categories of staff. According to the Frascati Manual, which constitutes the conceptual and methodological basis for OECD R&D surveys, the number of persons working in R&D should be expressed in *"full-time equivalent"* (or FTE). If need be, the Japanese practice might be accepted for purposes of international comparisons as far as the Government, Business Enterprise and PNP series are concerned where one may assume that the majority of the people concerned are essentially engaged in R&D; this is less the case as far as the Higher Education sector is concerned. According to an OECD study covering some ten Member countries of varying size, for which data are available, it appears that, on average, university professors engaged in R&D do not usually spend more than half their working time on this activity, independently of fields of science. The remaining half of the time is spent on preparing courses, on teaching, participation at meetings and conferences and various administrative activities. For the above reasons, it was considered absolutely necessary to adjust the R&D data for the Higher Education sector in Japan (and this adjustment will necessarily have an impact on the data at total national level).
The conditions and proportions of the adjustments of the R&D data are as follows:

Higher education sector: source of funds and numbers of researchers, 1981

	Official data	Adjustment co-efficient	Adjusted data
Source of funds (billion yen)			
Business Enterprise Sector	14.1	1.0	14.1
Public Sector:			
– Direct Government	163.5	1.0	163.5
– General University Funds	671.1	0.6	402.7
– Total public	834.6	–	566.2
Higher Education Sector	596.2	0.6	357.7
PNP Sector	0.7	1.0	0.7
Funds from Abroad	0.1	1.0	0.1
Total R&D Expenditures in the Higher Education Sector	1445.7	–	938.8
Number of Researchers	104 112	0.5	52 056

An average co-efficient of 0.6 is used to adjust the general university funds and the funds of the Higher Education sector, independently of the fields of science concerned (natural sciences and engineering, medical sciences, agricultural sciences, social sciences and humanities). The other sources of funds are considered as wholly devoted to R&D. An average co-efficient of 0.5 is used to adjust the number of researchers, independently of fields of science.

5. Total government expenditure covers three National Accounts classes for which data are available for practically all OECD countries, i.e. government final consumption, government gross fixed capital formation and subsidies.

6. These broad goals encompass several distinct objectives. Thus, "Economic Development" covers both agriculture and industry (i.e. development of agriculture, forestry and fishing, plus promotion of industrial development) and energy and infrastructure (production and national use of energy; transport and telecommunications; urban and rural planning; exploration and exploitation of the earth and the atmosphere). "Health and Welfare" covers health, protection of the environment, and social development and services. "Advancement of Knowledge" includes public general university funds (GUF), as well as other non-oriented civil R&D funds sometimes referred to as "Advancement of Research".

7. These observations are based on work carried out on indicators of government financial support for industry-related R&D.

8. Japan has recently abandoned this system, but this is not yet reflected in the data.

9. A statistical test of the assumption that the proportion of patents granted is higher for foreign applications than for domestic applications led to the conclusion that the former are on average of "better quality".

10. Germany, Austria, the United States, Finland, France, Italy, Japan, the Netherlands, the United Kingdom.

11. The TBP data used are those of the Statistics Bureau, Management and Coordination Agency. These data differ from those published by the Bank of Japan.
12. These observations are based on monographs elaborated by OECD which give, by country, the principal qualitative and quantitative aspects of the technological balance of payments.
13. For Germany, only payments are comparable with the data from other countries.
14. As we shall see later, industrial and export performance in R&D intensive activities also depend on general factors involving industrial competitiveness.
15. The Measurement of Scientific and Technical Activities – Proposed Standard Practice for Surveys of Research and Experimental Development "Frascati Manual" 1980 (OECD, 1981).
16. United States, Japan, Germany, France, United Kingdom, Italy, Canada, Australia, the Netherlands, Sweden and Belgium.
17. The preceding analysis has not taken the technological variable into account. The introduction of R&D intensities as endogenous variables does not alter the results.
18. The relative propensity to export (to perform R&D) is equal to the ratio of the weight of each industry category in exports (in R&D) to its weight in output. For example:

relative propensity to export of category $i = \dfrac{X_i}{X_m} \Big/ \dfrac{P_i}{P_m}$

where: X : exports
P : output
i : category
m : manufacturing total

19. Its trend can be measured by the ratio $\dfrac{X_i/X}{P_i/P}$

($\dfrac{X_i}{X}$, $\dfrac{P_i}{P}$, Weight of high-intensity industries in exports and output).

20. As in the formula: $CA_{ij} = 100 \times \dfrac{X_{ij}}{\sum_i X_{ij}} \Big/ \dfrac{X_{im}}{\sum_i X_{im}}$

X_{ij} : Exports of industry j in country i
X_{im} : Manufacturing exports of country i

21. Any assessment in absolute terms would call for the consideration of relative price levels – and not price trends – starting from a single reference point (1970).
22. This indicator, denoted IR,
is equal to ER + (1 – ER) MP
or to MP + (1 – MP) ER
where ER = proportion of output exported
MP = penetration rate.

TECHNICAL ANNEX

DETAILED SUPPORTING TABLES AND GRAPHS

Table 1. **Gross domestic expenditure on R&D (GERD): national trends**
All fields of science

	1981		Compound real growth rates (%)					1983
	Million $	Percentage of total[b]	1969-81	1969-75	1975-81	1979-81	1981-83	Million $
United States	73 678.0	46.4	1.8	−0.6	4.2	4.5	3.8	88 329.0
Japan[c]	25 574.5	16.1	8.1	8.3	7.9	11.2	8.2	33 493.7
Germany	15 644.8	9.9	5.4	6.2	4.7	2.6	1.9	18 130.2
France	10 700.8	6.7	3.1	2.3	4.2	5.0	4.7	13 134.4
United Kingdom[a]	11 369.8	7.2	2.2	1.3	3.1	2.9	−0.7	12 552.8
Italy	4 546.6	2.9	4.7	4.9	4.6	11.4	4.9	5 568.0
Canada	3 877.3	2.4	1.9	−0.9	4.8	8.5	3.5	4 619.2
Spain	908.0	0.6	9.4	15.7	3.4	3.8	..	..
Australia	1 539.5	1.0	..	..	2.2	2.8	..	..
Netherlands	2 508.4	1.6	2.3	3.6	0.9	0.1	3.5	2 992.4
Turkey	..	..	..	..	..	..	..	..
Sweden (n.s.e.)	2 166.7	1.4	7.2	8.9	5.6	10.0	7.1	2 776.8
Belgium[d]	1 065.7	0.9	4.3	4.4	4.1	..	..	..
Switzerland	1 785.4	1.1	1.2	1.6	0.8	0.6	−0.5	1 979.8
Austria	765.2	0.5	9.9	12.9	6.9	6.9	4.5	932.9
Yugoslavia	521.5	..	0.3	..	..	..	..	..
Denmark	539.8	0.3	3.1	3.4	2.9	4.9	..	..
Norway	593.0	0.4	6.3	9.3	3.3	−0.4	6.4	757.6
Greece	102.1	0.1	..	..	..	..	..	..
Finland[e]	499.3	0.3	8.2	8.7	7.7	10.5	8.5	655.2
Portugal[e]	154.5	0.1	3.7	0.7	6.9	6.1	..	..
New Zealand[a]	240.1	0.2	7.0	11.7	2.6	8.9	..	..
Ireland	155.3	0.1	5.1	7.6	2.7	5.9	..	..
Iceland	18.0	0.0	9.9	19.9	0.7	4.0	..	..
Total OECD[a]	158 720	100	3½	2	4½	5½	4	..
of which EEC[a]	46 940	29½	4	3½	4	4	2½	..

Source: OECD/STIIU data bank - December 1985.
a) OECD estimates.
b) Secretariat estimates are provided where 1981 figures are not available.
c) Data adjusted by OECD would be 23 408 million $ in 1981 or 15 % of the OECD total, and in 1983, 30 750 million $.
d) 1979.
e) 1982.

Table 2. **Number of researchers: national trends**
All fields of science

	1981		Compound real growth rates (%)					
	Full-time equivalent	Percentage of total[b]	1969-81	1969-75	1975-81	1979-81	1981-82	1982-83
United States	683 700	41.8	1.8	−0.8	4.4	5.5	2.7	2.5
Japan[c]	392 625	24.0	5.0	6.4	3.6	3.9	3.4	7.2
Germany	128 162	7.8	4.6	5.6	3.6	2.5	..	..
France	85 500	5.2	3.4	2.2	4.6	8.3	5.4	2.9
United Kingdom (n.s.e.)[d]	104 445	6.4	..	..	..	..	..	..
Italy	52 060	3.2	6.2	6.9	5.4	5.9	8.9	11.1
Canada	29 976	1.8	3.3	1.8	4.9	6.9	6.6	2.6
Spain	14 376	0.9	8.9	10.2	7.5	3.4	..	..
Australia	24 210	1.5	−1.0	−2.7	0.7	2.8	..	..
Netherlands	19 436	1.2	3.5	3.1	3.9	2.6	..	..
Turkey	..	..	..	..	..	..	..	..
Sweden (n.s.e.)	15 035	0.9	6.8	7.7	5.9	13.1	..	..
Belgium[e]	10 943	0.8	3.5	1.2	5.9	..	..	..
Switzerland[e]	10 720	0.6	2.2	6.0	−1.4	..	..	..
Austria	6 712	0.4	5.2	6.7	3.7	3.7	..	..
Yugoslavia	24 881	1.5	..	..	..	..	..	..
Denmark[e]	5 988	0.4	3.6	2.6	4.6	..	..	..
Norway	7 496	0.5	7.9	11.6	4.3	2.9	..	..
Greece[e]	2 634	0.2	..	..	..	..	..	..
Finland[g]	9 421	0.5	4.1	2.7	5.6	..	..	..
Portugal[f]	3 019	0.2	2.4	−2.0	7.0	10.0	..	..
New Zealand	..	..	..	..	..	..	..	..
Ireland	2 636	0.2	4.5	8.1	1.1	0.3	5.2	..
Iceland	345	0.0	10.4	12.7	8.2	7.0	..	..
Total OECD[a]	1 634 880	100	3	2	4	4½	..	..
of which EEC[a]	414 640	25½	3	3	3	3½	..	..

Source: OECD/STIIU data bank - December 1985.
a) OECD estimates.
b) Secretariat estimates are provided where 1981 figures are not available.
c) Not in full-time equivalent. Secretariat estimate of f.t.e. is 311 000.
d) 1978.
e) 1979.
f) 1982.
g) 1983.

Table 3. **Government-financed GERD in OECD countries**
All fields of science

	1981		Compound real growth rates (%)					1983
	Million $	Percentage of total [b]	1969-81	1969-75	1975-81	1979-81	1981-83	Million $
United States	36 281.0	51.0	0.1	-2.2	2.4	1.9	3.6	43 386.0
Japan	6 891.6	9.7	7.1	8.0	6.2	6.5	2.1	8 028.1
Germany	6 513.7	9.2	5.4	8.4	2.5	1.6	1.0	7 420.7
France	5 654.3	8.0	1.7	-0.2	3.6	7.1	7.5	7 086.5
United Kingdom [a]	5 575.5	7.8	1.8	2.4	1.2	3.0	0.5	6 304.0
Italy	2 146.2	3.0	5.4	4.6	6.2	15.5	10.5	2 917.9
Canada	1 917.2	2.7	0.4	-0.2	1.0	2.1	6.4	2 411.6
Spain	479.9	0.7	9.9	12.7	7.2	6.6	..	..
Australia	1 167.3	1.6	..	..	..	..	..	..
Netherlands	1 184.9	1.7	3.0	4.2	1.8	-0.8	..	..
Turkey	..	..	..	..	..	..	..	..
Sweden (n.s.e.)	865.0	1.2	7.2	8.4	6.0	12.9	2.5	1 015.7
Belgium [c]	330.4	0.5	-2.9	-0.9	-4.8	..	..	..
Switzerland	..	0.6	5.5	5.4	5.6	-1.2	-1.2	447.0
Austria	335.3	0.5	9.2	14.7	4.1	4.1	..	..
Yugoslavia	188.3	0.3	..	..	..	..	..	..
Denmark [c]	214.8	0.4	3.0	3.4	2.6	..	..	..
Norway	339.0	0.5	6.2	9.8	2.8	-2.6	..	..
Greece	86.2	0.1	..	..	..	..	..	..
Finland	229.8	0.3	7.6	8.5	6.8	11.2	3.9	277.0
Portugal [d]	95.7	0.1	3.6	1.9	5.4	-1.8	..	..
New Zealand [a]	196.4	0.3	4.7	6.8	2.6	..	..	..
Ireland	87.7	0.1	5.0	8.7	1.4	5.9	..	..
Iceland	15.4	0.0	9.1	17.9	0.9	6.6	..	..
Total OECD [a]	71 110	100	1½	½	3	3½	..	..
of which EEC [a]	21 890	31	3	3½	2½	4½	..	..

Source: OECD/STIIU data bank - December 1985.
a) OECD estimates.
b) Secretariat estimates are provided where 1981 figures are not available.
c) 1979.
d) 1982.

Table 4. **Business-financed GERD in OECD countries**
All fields of science

	1981		Compound real growth rates (%)					1983
	Million $	% of total [b]	1969-81	1969-75	1975-81	1979-81	1981-83	Million $
United States	35 944.0	44.7	4.1	1.8	6.4	7.4	4.0	43 246.0
Japan	15 928.8	19.8	8.6	7.8	9.3	14.4	10.7	21 822.1
Germany	8 918.6	11.1	5.4	3.8	7.0	4.1	2.9	10 540.8
France	4 367.6	5.4	5.5	5.5	5.4	3.4	6.2	5 510.7
United Kingdom [a]	4 695.1	5.8	2.0	-0.4	4.4	2.1	0.3	5 282.5
Italy	2 277.1	2.8	5.0	5.6	4.3	6.7	-0.5	2 508.7
Canada	1 628.1	2.0	4.9	-0.9	11.0	17.6	-0.8	1 781.4
Spain	416.4	0.5	9.7	19.0	1.0	2.0	..	..
Australia	323.0	0.4	..	..	..	..	..	..
Netherlands	1 161.9	1.4	1.0	2.2	-0.3	-1.0	..	..
Turkey	..	..	..	..	..	..	..	..
Sweden (n.s.e.)	1 241.9	1.5	7.3	9.0	5.7	7.2	10.3	1 686.9
Belgium [c]	700.9	1.2	8.3	9.9	6.8	..	..	..
Switzerland	1 220.0	1.5	0.5	1.0	0.0	-3.8	5.9	1 532.8
Austria	384.4	0.5	9.6	11.3	8.0	8.0	3.2	457.0
Yugoslavia	298.1	0.4	..	..	..	..	..	..
Denmark [c]	179.6	0.3	3.3	3.1	3.6	..	..	..
Norway	237.5	0.3	6.3	8.0	4.6	2.6	..	..
Greece	15.9	0.0	..	..	..	..	..	..
Finland	259.1	0.3	9.3	10.1	8.6	9.7	12.2	364.2
Portugal [d]	46.3	0.0	3.7	-3.4	11.2	31.8	..	..
New Zealand [a]	43.5	0.1	7.8	12.3	3.5	..	..	..
Ireland	58.6	0.1	5.4	5.2	5.6	9.8	..	..
Iceland	1.0	0.0	11.7	24.1	0.5	2.8	..	..
Total OECD [a]	80 360	100	5	3½	6½	7½	..	..
of which EEC [a]	22 700	28	4½	3½	5½	3½	..	..

Source: OECD/STIIU data bank - December 1985.
a) OECD estimates.
b) Secretariat estimates are provided where 1981 figures are not available.
c) 1979.
d) 1982.

Table 5. **Recent growth of public R&D funding by major socio-economic objective**

	Agriculture and industry		Energy and infrastructure		Health and welfare		Advancement of knowledge		Defence and space		Total public R&D funding	
	1981-83	1983-84	1981-83	1983-84	1981-83	1983-84	1981-83	1983-84	1981-83	1983-84	1981-83	1983-84
United States	−2.6	−1.9	−15.0	1.7	−2.7	8.7	0.3	10.2	6.3	12.4	1.6	10.3
Japan[a]	..	..	..	..	−4.1	..	5.4	..	2.3	..	2.1	..
Germany	5.8	−1.7	−5.9	1.4	−8.5	1.1	1.8	2.1	2.2	1.9	−0.2	1.3
France	16.8	11.6	2.1	9.6	−1.7	4.2	10.7	4.5	−1.6	4.7	4.8	5.8
United Kingdom	2.8	16.2	−6.8	−0.4	−1.9	5.0	0.6	−0.3	0.1	7.4	−0.2	6.2
Italy	16.7	−4.4	−1.2	−2.6	17.3	−1.7	−2.0	−3.4	−6.1	66.1	3.7	3.6
Canada	4.2	..	7.5	..	2.3	..	7.4	..	8.9	..	6.1	..
Spain	−8.0	..	−2.9	..	21.8	..	0.1	..	5.5	..	−0.6	..
Australia	−1.6	..	6.6	..	10.3	..	2.6	..	4.3	..	2.5	..
Netherlands	3.4	−45.4	−12.0	5.6	−5.8	1.6	−2.1	18.2	−23.4	−11.4	−3.3	1.4
Turkey	−31.8	−31.8	−31.0	−29.5	−24.3	−23.4	3.0	3.0	..	..	−16.7	−16.7
Sweden	3.2	47.6	−1.0	−2.9	−3.2	−11.9	3.5	1.6	35.2	0.7	8.6	1.6
Belgium	−6.4	2.4	−14.6	6.5	−29.9	0.0	25.8	−1.0	6.0	14.4	−1.9	0.9
Switzerland	6.3	..	3.9	..	0.1	..	−2.4	..	−16.2	..	−5.9	..
Austria	0.4	4.9	−3.4	6.7	6.6	0.2	5.5	1.3	−31.4	239.1	4.4	2.0
Yugoslavia	..	..	..	..	..	..	..	..	..	..	..	..
Denmark	0.8	4.6	−0.1	−20.2	−30.8	2.6	23.4	0.4	6.4	0.7	4.7	−1.3
Norway	−0.7	..	−2.8	..	−2.1	..	4.2	..	21.8	..	2.6	..
Greece	−19.8	−16.5	30.0	87.7	26.2	30.2	5.4	5.4	−5.1	−4.8	2.8	−38.1
Finland	5.1	9.9	7.5	−5.6	10.4	27.6	6.1	−0.1	7.1	6.6	6.4	5.1
Portugal	16.0	..	7.0	..	27.6	..	19.9	..	..	..	9.0	..
New Zealand	0.7	..	−1.4	..	7.3	..	−1.5	..	..	..	0.4	..
Ireland	6.4	−2.3	5.1	−8.9	10.6	−0.3	−9.9	−4.3	−9.4	19.0	2.2	−2.9
Iceland	..	28.2	..	45.7	..	22.4	..	57.7	..	..	..	39.0
Total OECD[a]	5	2½	−7	2½	−2½	6	3½	3½	4½	11	2	6½

Source: OECD/STIIU data bank - December 1985.
a) OECD estimates.

Table 5 *(cont'd)*

Composition of the major socio-economic objectives

Major objectives	OECD socio-economic objectives
Agriculture and industry:	"Development of agriculture, forestry and fishing" + "Promotion of industrial development"
Energy and infrastructure	"Production and rational use of energy" + "Transport and telecommunications" + "Urban and rural planning" + "Exploration and exploitation of earth and atmosphere
Health and welfare	"Health" + "Protection of the environment" + "Social development and services"
Advancement of knowledge	"Advancement of knowledge" including public general university funds
Defence and space:	"Defence" + "Civil space"

Table 6. **GERD by Sector of performance and Source of funds - 1983**
All fields of science

	Percentage by Sector of performance				Total	Percentage by Source of funds			
	Business enterprise	Government	Higher education	Private non-profit		Business enterprise	Public	Other national	Funds from abroad
United States	71.1	12.4	13.4	3.0	100.0	49.0	48.4	2.7	0.0
Japan	63.5	9.6	23.0	3.9	100.0	65.2	24.0	10.8	0.1
Germany[c]	68.3	14.3	16.8	0.5	100.0	57.0	41.6	0.4	0.9
France	56.8	26.4	15.8	0.9	100.0	42.0	54.0	0.5	3.6
United Kingdom[a]	61.3	21.7	14.0	3.0	100.0	42.0	49.5	3.0	5.5
Italy	53.5	..	..	..	100.0	42.5	55.4	..	2.2
Canada	46.9	26.9	25.0	1.2	100.0	38.6	52.2	5.0	4.2
Spain[c]	48.8	33.8	17.4	..	100.0	45.9	52.9	0.1	1.2
Australia[c]	22.4	46.5	29.7	1.4	100.0	21.0	75.8	2.1	1.1
Netherlands[c]	53.3	20.8	23.2	2.8	100.0	46.3	47.2	1.3	5.2
Turkey	..	..	..	..	..	..	..	..	..
Sweden (n.s.e.)	67.5	5.3	27.0	0.2	100.0	60.7	36.6	1.1	1.5
Belgium[b]	69.6	9.4	20.6	0.4	100.0	65.8	31.0	1.8	1.4
Switzerland[c]	74.2	5.9	19.9	..	100.0	68.3	21.4	..	0.0
Austria[c]	55.8	6.0	32.8	2.3	100.0	50.2	43.8	0.4	2.5
Yugoslavia[c]	56.4	24.6	18.9	..	100.0	57.2	36.1	..	1.8
Denmark[a,d]	51.8	21.8	25.6	0.8	100.0	44.8	51.1	2.0	2.1
Norway[c]	52.1	18.4	29.0	0.5	100.0	40.1	57.2	1.4	1.4
Greece[c]	22.5	63.1	14.5	..	100.0	15.6	84.4	..	..
Finland[c]	57.1	25.8	16.5	0.6	100.0	54.2	43.6	1.2	1.0
Portugal[d]	31.2	43.6	20.6	4.6	100.0	30.0	61.9	4.8	3.3
New Zealand[b]	20.0	60.3	16.9	2.8	100.0	15.7	84.2	0.1	0.0
Ireland[c]	43.6	39.3	16.0	1.1	100.0	37.7	56.5	1.1	4.8
Iceland[c]	9.6	60.7	26.0	3.7	100.0	5.7	85.6	4.4	4.3

Source: OECD/STIIU data bank - December 1985.
a) OECD estimates.
b) 1979.
c) 1981.
d) 1982.

Table 7. **Business enterprise R&D expenditure (BERD): national trends**

	1981		Compound real growth rates (%)					1983
	Million $	Percentage of total[b]	1969-81	1969-75	1975-81	1979-81	1981-83	Million $
United States	51 810.0	50.2	1.9	-1.6	5.4	6.5	4.4	62 816.0
Japan	15 517.3	15.0	8.4	7.6	9.2	13.6	10.7	21 270.0
Germany	10 686.3	10.4	5.9	5.6	6.2	2.0	2.9	12 649.0
France	6 304.4	6.1	4.2	3.9	4.4	5.7	2.9	7 461.1
United Kingdom (n.s.e.)	7 029.7	6.8	2.0	-0.1	4.1	2.2	-1.3	7 662.9
Italy	2 563.1	2.5	5.6	6.3	4.8	9.5	5.5	3 179.0
Canada	1 902.9	1.8	5.9	1.9	10.1	16.9	1.3	2 168.0
Spain	442.7	0.4	10.9	21.9	0.9	2.5	..	..
Australia	344.4	0.3	..	..	1.6	1.4	..	..
Netherlands	1 336.0	1.3	1.7	2.7	0.8	1.7	3.9	1 607.2
Turkey	..	..	..	..	..	..	..	..
Sweden (n.s.e.)	1 442.1	1.4	6.8	8.5	5.1	7.5	7.9	1 874.6
Belgium	950.1	0.9	7.0	8.5	5.4	3.6	4.1	1 148.7
Switzerland	1 324.8	1.3	0.5	0.7	0.3	-0.1	-0.4	1 471.0
Austria	427.4	0.4	11.0	13.4	8.6	5.9	..	..
Yugoslavia	294.4	0.3	..	..	..	..	..	..
Denmark	273.8	0.3	3.7	1.9	5.5	5.2	10.7	372.3
Norway	309.1	0.3	7.3	10.0	4.7	2.4	9.0	414.2
Greece	22.9	0.0	..	..	..	..	..	..
Finland	272.9	0.3	9.2	9.9	8.6	10.4	10.6	372.3
Portugal[c]	48.3	0.0	5.6	-0.8	12.5	31.7	..	..
New Zealand[a]	52.0	0.1	7.2	12.2	2.5	..	..	..
Ireland	67.7	0.1	7.2	5.5	9.0	14.9	..	..
Iceland	1.7	0.0	46.4	73.8	23.2	5.8	..	..
Total OECD[a]	103 120	100	3½	1	5½	6½	..	..
of which EEC[a]	29 230	28½	4	3½	5	3½	..	..

Source: OECD/STIIU data bank - December 1985.
a) OECD estimates.
b) Secretariat estimates are provided where 1981 figures are not available.
c) 1982.

Table 8. **R&D intensity[i] and profit-earning capacity[ii] of manufacturing industry**

	R&D expenditure as a percentage of value added					Gross operating surplus as a percentage of value added					
	1969	1975	1979	1981	1983	1969	1975	1981	1983	1969-75 average	1975-83 average
United States	7.4	6.9	6.8	8.1	9.3	25.2	25.2	22.7	21.2	24.3	24.3
Japan	2.9	3.8	4.2	4.9	5.7	56.7	41.3	43.9	42.6	51.3	43.3
Germany	3.3	4.1	4.9	5.4	..	36.7	29.1	25.6	26.9	32.2	28.3
France	3.4	4.0	4.3	..	..	35.9	29.1	30.1	..	34.0	29.5
United Kingdom	4.7	4.4	4.8	6.6	..	26.8	17.7	20.3	25.1	24.1	21.6
Italy	1.6	1.5	1.4	1.7	1.8	34.1	27.7	34.5	32.8	32.0	33.1
Canada	2.1	1.3	1.9	2.6	3.1	32.0	32.8	32.9	27.5	32.5	30.9
Australia	..	0.8	0.9	0.8	..	33.7	25.2	25.4	23.5	29.4	25.5
Netherlands	4.1	4.0	4.8	5.6	..	..	..	22.0	..	..	..
Sweden	2.9	4.1	5.5	6.3	7.4	28.9	26.7	20.6	31.0	26.5	21.6
Belgium	2.5	3.4	3.8	4.2	..	36.3	21.4	19.2	22.4	31.1	21.0
Denmark	2.1	2.1	2.4	2.7	..	27.8	25.5	29.0	32.2	25.1	26.0
Norway	2.0	2.4	2.7	3.0	..	29.3	30.4	26.2	27.2	30.4	27.4
Finland	1.4	1.8	2.1	2.5	2.9	42.5	31.9	35.4	38.1	36.7	34.7
Portugal	..	..	0.1	..	..	..	30.8	38.1	..	..	..
New Zealand	..	0.8	0.7	..	..	39.5	33.0	36.4	34.6	30.0	35.2

Source: OECD/STIIU data bank - December 1985.
Note: Nearest year depending on availability.
i) Measured by R&D expenditure as a percentage of value added.
ii) Measured by gross operating surplus as a percentage of value added.

Table 9. **Comparative structure of BERD[a] by industry group[b] - 1981**
when BERD = 100

	Electrical	Chemical	Aerospace	Other transport	Basic metals	Machinery	Chemical-linked	Other manu-facturing	Services	BERD[c]
United States	20.2	13.9	22.6	10.8	2.9	20.2	3.1	3.1	4.1	100.0
Japan	24.5	18.1	0.0	17.2	8.3	13.3	7.2	4.2	6.6	100.0
Germany	23.9	23.1	6.2	14.1	4.5	16.1	3.2	1.9	2.4	100.0
France	24.7	18.8	17.5	11.8	3.3	9.2	5.3	2.1	5.8	100.0
United Kingdom	31.1	16.1	20.1	5.0	2.4	12.0	4.7	2.1	4.8	100.0
Italy	14.9	23.2	9.1	14.4	2.5	10.0	3.9	4.2	16.5	100.0
Canada	22.5	18.0	12.3	2.4	6.0	7.3	3.9	5.1	12.2	100.0
Spain	16.3	22.8	37.6	19.6	6.9	5.5	7.2	5.1	14.1	100.0
Australia	10.9	15.6	..	9.4	10.0	5.8	5.8	3.9	32.0	100.0
Netherlands	..	34.2	..	..	..	..	7.2	0.7	7.5	100.0
Turkey	..	..	..	..	..	..	..	..	..	..
Sweden	23.1	9.8	..	21.9	7.1	14.7	3.5	6.4	11.1	100.0
Belgium	25.1	34.0	0.4	2.5	8.1	5.9	6.3	5.0	12.2	100.0
Switzerland	24.7	48.5	..	0.6	4.8	16.0	2.6	0.8	2.0	100.0
Austria	22.8	12.2	..	9.3	8.9	23.8	9.3	5.4	7.2	100.0
Yugoslavia	..	..	..	..	..	..	..	..	..	..
Denmark	11.8	18.1	..	3.3	2.0	22.9	8.6	13.7	19.5	100.0
Norway	20.1	9.4	..	5.1	11.4	15.2	4.6	4.3	11.2	100.0
Greece	5.9	20.3	8.0	5.8	12.4	0.9	4.6	11.6	30.6	100.0
Finland	20.9	16.1	0.2	3.0	8.2	24.6	7.8	12.0	5.7	100.0
Portugal	16.8	22.8	0.0	7.9	2.9	3.3	3.9	4.2	29.0	100.0
New Zealand	9.4	11.2	0.2	4.2	5.8	3.3	24.0	5.0	34.4	100.0
Ireland	22.6	17.2	0.2	3.3	6.0	8.1	26.6	5.8	8.5	100.0
Iceland	31.8	13.0	..	..	27.3	1.3	8.4	7.1	0.1	100.0
Total OECD[b]	22	17	15	11½	4	17	4	3	5½	100

Source: OECD/STIIU data bank - December 1985.
a) BERD = R&D performed in the Business Enterprise sector.
b) Partly estimated by OECD, except for the Netherlands.
c) The sum of the indicated industry groups may be less than 100 %, the difference representing R&D in agriculture and mining.

Table 9 (cont'd). **Detailed composition of industry groups**

Electrical group:	Electrical machinery, electronic equipment and components (computers excluded)
Chemical group:	Chemicals, drugs, petroleum refineries
Aerospace:	Aerospace (missiles included)
Other transport:	Motor vehicles, ships, other transport
Basic metals group:	Ferrous metals, non-ferrous, fabricated metal products
Machinery group:	Instruments, office machinery and computers, machinery n.e.c.
Chemical linked group:	Food, drink and tobacco; textiles and clothing; rubber and plastics
Other manufacturing group:	Stone, clay and glass; paper and printing; wood, cork and furniture; other manufacturing
Services group:	Utilities, construction, transport and storage, communications, commerce and engineering services, other services

Table 10. **Country positions in the different industry groups**[a] **- 1981**
when OECD=100

	Electrical	Chemical	Aerospace	Other transport	Basic metals	Machinery	Chemical-linked	Other manu-facturing	Services	BERD
United States	46.1	40.9	75.3	47.2	35.7	62.0	37.1	49.6	38.8	50.2
Japan	16.9	16.0	0.0	22.7	30.3	12.3	25.9	19.8	18.5	15.1
Germany	11.3	14.0	4.3	12.7	11.4	10.2	7.9	6.3	4.6	10.4
France	6.9	6.8	7.2	6.3	4.9	3.5	7.8	4.0	6.7	6.2
United Kingdom	9.7	6.5	9.2	3.0	4.0	5.0	7.7	4.5	6.1	6.9
Italy	1.7	3.4	1.5	3.1	1.5	1.5	2.3	3.3	7.7	2.5
Canada	1.8	1.8	1.4	0.4	2.5	0.8	1.6	2.8	4.0	1.7
Spain	0.3	0.6	1.1	0.7	0.7	0.1	0.7	0.7	1.1	0.4
Australia	0.2	0.3	..	0.3	0.8	0.1	0.5	0.4	2.0	0.3
Netherlands	..	2.6	..	..	..	..	2.2	0.3	1.8	1.3
Turkey	..	..	..	..	..	..	..	..	..	..
Sweden	1.5	0.8	..	2.7	2.4	1.3	1.2	2.8	2.9	1.4
Belgium	1.1	1.9	0.0	0.2	1.8	0.3	1.4	1.5	2.1	0.9
Switzerland	1.3	3.4	..	0.1	1.4	1.2	0.7	0.3	0.5	1.2
Austria	0.4	0.3	..	0.3	0.9	0.6	0.9	0.7	0.6	0.4
Yugoslavia	..	..	..	..	..	..	..	..	..	..
Denmark	0.1	0.3	..	0.1	0.1	0.4	0.5	1.1	1.0	0.3
Norway	0.3	0.2	..	0.1	0.8	0.3	0.3	0.4	0.6	0.3
Greece	0.0	0.0	0.0	0.0	0.1	0.0	0.0	0.1	0.1	0.0
Finland	0.3	0.3	0.0	0.1	0.5	0.4	0.5	1.0	0.3	0.3
Portugal	0.0	0.1	0.0	0.0	0.0	0.0	0.0	0.1	0.2	0.0
New Zealand	0.0	0.0	0.0	0.0	0.1	0.0	0.3	0.1	0.3	0.0
Ireland	0.1	0.1	0.0	0.0	0.1	0.0	0.4	0.1	0.1	0.1
Iceland	0.0	0.0	..	..	0.0	0.0	0.0	0.0	0.0	0.0
Total OECD	100	100	100	100	100	100	100	100	100	100

Source: OECD/STIIU data bank - December 1985.
a) Partly estimated by OECD, except for the Netherlands.

Table 11. **Industry group: Electrical-electronic - 1981**

	Million $	Researchers	Expenditure					
			Composition (%)		% Financed by		Compound growth rates	
			Electric	Electronic	Private	Public	1975-79	1979-81
United States	10 329.0	109 700	33.0	67.0	62.0	38.0	4.0	5.1
Japan	3 808.2	55 961	31.9	68.1	98.7	1.4	8.4	17.1
Germany	2 559.2	28 997	..	..	84.7	12.6	6.2	-0.6
France	1 555.5	9 713	14.2	85.8	73.6	26.4	3.1	8.7
United Kingdom	2 189.3	29 108	10.2	89.8	..	45.6	..	..
Italy	382.0	4 079	23.9	76.1	80.5	18.1	5.6	-4.3
Canada	416.2	3 887	16.8	83.2	82.8	11.4	-0.5	18.3
Spain	72.0	655	41.8	58.2	87.7	3.3	..	-4.6
Australia	37.6	324	28.0	72.0	85.2	..	..	..
Netherlands	..	..	..	..	..	..	..	..
Turkey	..	..	..	..	..	..	..	..
Sweden	333.8	2 165	..	..	86.9	11.4	6.9	9.8
Belgium	238.7	1 514	40.7	59.3	93.7	6.2	9.3	9.9
Switzerland[c]	301.4	1 500	68.4	31.6	..	..	..	-6.9
Austria	97.4	699	43.0	57.0	98.9	1.1	..	..
Yugoslavia	..	..	..	..	..	..	..	..
Denmark	32.4	261	29.4	70.6	91.9	6.8	3.5	9.7
Norway	61.3	545	27.2	72.8	87.4	11.3	7.1	-11.2
Greece	1.4	..	100.0	..	100.0	0.0	..	..
Finland	56.9	499	45.1	54.9	87.4	9.0	8.7	10.3
Portugal[e]	8.8	112	..	..	..	..	..	..
New Zealand[d]	3.7	..	..	..	..	..	-3.8	..
Ireland	15.2	167	25.4	74.6	66.3	19.3	10.2	46.3
Iceland	0.6	12	100.0	0.0	42.9	46.9	..	50.4
Total OECD[a,b]	22 500	262 910	24½	75½	75	24½	6	7

Source: OECD/STIIU data bank - December 1985.
a) OECD estimates.
b) At total level, structure percentage (industrial composition and sources of funds) consists of averages of those countries for which data are available. For the United States, industrial composition refers rather to that of 1980, totally or partially. For France, the breakdown by sources of funds relates to the total (intramural and extramural) R&D expenditure of the group. The sum of the two principal sources of funds, private and public, may be less than 100%, the difference representing other national sources and funds from abroad. For a more explicit description of the industry group components, see table 9.
c) Researchers for 1979.
d) Expenditure for 1979.
e) 1982.
Note: Computers are not included in this group, but in the machinery group.

Table 12. **Industry group: Chemical - 1981**

	Million $	Researchers	Expenditure						
			Composition (%)			% Financed by		Compound growth rates	
			Chemicals	Drugs	Petroleum refineries	Private	Public	1975-79	1979-81
United States[e]	6 188.0	77 200	46.0	29.0	25.0	91.5	8.5	4.3	..
Japan	2 812.6	35 526	60.6	33.2	6.2	98.7	0.9	6.0	9.9
Germany	2 469.8	11 502	..	..	3.3	94.8	4.7	3.3	4.4
France	1 187.4	5 838	49.4	32.4	18.1	91.3	4.9	4.3	7.6
United Kingdom	1 213.5	11 245	49.1	45.2	5.7	..	..	..	..
Italy	594.2	5 234	34.8	54.7	10.4	93.5	4.4	0.1	15.0
Canada	325.8	2 187	28.3	14.6	57.1	90.4	5.2	8.5	15.7
Spain	101.1	817	..	..	13.3	98.2	1.7	..	2.2
Australia	53.7	659	74.6	25.4	..	87.2	9.0	..	..
Netherlands	456.3	2 400	..	..	..	..	..	0.7	3.7
Turkey	..	..	..	..	..	..	..	..	..
Sweden	141.2	1 589	31.1	68.9	..	97.1	2.6	11.0	0.7
Belgium	323.3	1 172	75.3	23.8	0.9	95.1	4.9	1.1	5.1
Switzerland[d]	591.7	2 500	12.5	87.5	..	..	..	..	1.8
Austria	52.1	281	59.6	37.1	3.2	90.7	3.9	..	..
Yugoslavia	..	..	..	..	..	..	..	..	..
Denmark	49.7	534	34.3	65.7	..	98.0	1.9	3.7	1.4
Norway	28.7	299	71.9	21.6	6.5	85.3	14.2	5.0	-7.7
Greece	4.6	..	..	..	..	100.0	0.0	..	..
Finland	43.9	522	44.5	41.9	13.6	95.2	4.8	5.9	18.8
Portugal[f]	8.2	95	..	..	..	..	..	..	..
New Zealand[c]	5.0	..	..	..	..	..	..	-4.1	..
Ireland	11.6	139	60.1	39.9	0.0	82.6	8.9	7.1	10.8
Iceland	0.2	5	100.0	0.0	0.0	100.0	0.0	61.3	-18.0
Total OECD[a,b]	17 690	159 640	75½	19½	5	95	3½	4	6½

Source: OECD/STIIU data bank - December 1985.
a) OECD estimates.
b) See explanation table 11.
c) Expenditure for 1979.
d) Researchers for 1979.
e) Expenditure for 1980.
f) 1982.

Table 13. **Industry group: Machinery - 1981**

	Million $	Researchers	Expenditure						
			Composition (%)			% Financed by		Compound growth rates	
			Instruments	Office machinery and computers	Machinery n.e.c	Private	Public	1975-79	1979-81
United States	10 432.0	118 700	33.0	44.0	22.0	87.3	12.8	6.4	9.2
Japan	2 068.6	30 994	26.2	23.8	50.0	98.6	0.9	11.7	15.9
Germany	1 715.9	15 548	12.1	..	..	89.8	9.7	15.9	0.0
France	581.0	4 820	13.8	50.1	36.1	83.8	6.3	1.3	4.9
United Kingdom	842.4	9 036	13.3	38.5	48.2	90.8	9.2	..	..
Italy	256.5	2 287	8.2	64.4	27.4	90.4	9.4	6.2	22.8
Canada	150.4	1 235	11.9	35.7	52.4	66.7	11.3	-0.3	21.9
Spain	24.4	166	9.0	42.6	48.3	91.3	8.0	..	22.1
Australia	19.9	238	25.9	..	74.1	88.3	..	..	..
Netherlands	..	..	..	..	..	..	..	..	..
Turkey	..	..	..	..	..	..	..	..	..
Sweden	212.3	1 525	12.2	..	..	91.5	7.7	1.3	2.2
Belgium	55.7	478	10.1	0.3	89.6	97.8	2.2	12.1	-0.9
Switzerland[c]	195.6	350	81.1	1.6	17.2	..	..	..	15.7
Austria	101.7	769	1.5	34.9	63.6	95.0	4.3	..	..
Yugoslavia	..	..	..	..	..	..	..	..	..
Denmark	62.6	434	33.4	..	..	89.1	6.9	3.7	6.0
Norway	46.5	442	2.8	23.3	74.0	76.1	22.8	5.7	1.6
Greece	0.2	..	..	..	100.0	100.0	0.0	..	..
Finland	67.0	614	18.1	13.4	68.5	85.5	14.1	13.3	11.3
Portugal[e]	1.8	14	..	..	..	..	..	..	..
New Zealand[d]	1.4	..	..	..	..	..	..	-6.9	..
Ireland	5.4	61	16.9	41.3	41.8	82.6	17.4	17.1	22.1
Iceland	0.0	1	100.0	0.0	0.0	50.0	50.0	..	..
Total OECD[a,b]	17 490	192 110	28½	8½	63	89	10½	5	8

Source: OECD/STIIU data bank - December 1985.
a) OECD estimates.
b) See explanation table 11.
c) Researchers for 1979.
d) 1979.
e) 1982.
Note: Computers are included in this group.

Table 14. **Aerospace - 1981**

	Million $	Researchers	Expenditure			
			% Financed by		Compound growth rates	
			Private	Public	1975-79	1979-81
United States	11 968.0	91 100	28.7	71.3	1.8	11.6
Japan	5.8	65	98.9	1.1	..	17.6
Germany	659.4	4 271	22.1	76.0	-1.9	4.8
France	1 104.1	5 362	18.2	68.4	2.8	4.1
United Kingdom	1 414.1	10 947	..	67.5	..	..
Italy	234.4	1 199	60.2	10.1	36.0	52.0
Canada	229.1	1 231	73.0	21.5	17.9	16.6
Spain[d]	9.0	87	..	..	..	..
Australia	..	..	..	..	..	..
Netherlands	..	..	..	..	..	..
Turkey	..	..	..	..	..	..
Sweden	..	..	..	..	..	..
Belgium	4.3	25	38.7	9.3	44.5	-20.1
Switzerland	0.0	..	..	..	..	..
Austria	0.0	0	..	..	..	..
Yugoslavia	..	..	..	..	..	..
Denmark	0.0	0	..	..	..	..
Norway	0.0	0	..	..	..	..
Greece	1.8	..	100.0	0.0	..	..
Finland	0.4	6	32.1	67.9	-16.8	11.2
Portugal[e]	0.0	0	..	..	..	..
New Zealand[c]	0.1	..	..	..	..	..
Ireland	0.2	2	100.0	0.0	-23.2	204.2
Iceland	0.0	0	..	..	..	..
Total OECD[a,b]	15 790	114 390	27½	70½	2	11

Source: OECD/STIIU data bank - December 1985.
a) OECD estimates.
b) See explanation table 11.
c) 1975
d) 1976.
e) 1982.

Table 15. **Industry group: Other transport - 1981**

	Million $	Researchers	Expenditure						
			Composition (%)			% Financed by		Compound growth rates	
			Motor vehicles	Ships	Other transport	Private	Public	1975-79	1979-81
United States[d]	5 117.0	31 200	97.0	..	..	86.0	14.0	10.1	..
Japan	2 676.4	18 093	83.6	15.2	1.1	95.4	4.6	6.2	15.5
Germany	1 501.9	7 275	98.2	1.1	0.7	95.3	3.3	13.5	6.7
France	742.4	2 420	96.6	0.8	2.5	98.6	1.1	10.3	0.5
United Kingdom	351.8	3 822	95.0	5.0	0.0	91.5	3.6	..	..
Italy	368.1	989	98.9	1.1	0.0	99.7	0.3	-3.9	-3.8
Canada	53.7	428	..	..	..	90.0[a]	10.0[a]	12.2	21.6
Spain	86.6	442	75.0	3.1	21.9	98.7	1.3	..	25.3
Australia	32.3	237	93.1	..	6.9	91.8	8.2	..	..
Netherlands	..	..	..	..	..	..	..	..	..
Turkey	..	..	..	..	..	..	..	..	..
Sweden	315.8	1 171	..	3.3	96.7	80.7	19.3	3.6	4.5
Belgium	23.9	77	82.2	7.7	10.1	93.2	6.8	35.8	-2.9
Switzerland	6.8	..	..	..	..	..	..	..	..
Austria	39.8	251	92.3	3.5	4.1	64.9	3.3	..	..
Yugoslavia	..	..	..	..	..	..	..	..	..
Denmark	9.2	80	0.0	93.4	6.6	95.2	4.7	-4.5	-1.1
Norway	15.5	125	11.7	79.7	8.5	76.6	20.1	-5.4	13.9
Greece	1.3	..	..	25.0	75.0	100.0	0.0	..	..
Finland	8.2	69	..	71.9	28.1	92.3	7.4	33.7	21.2
Portugal[e]	3.2	56	..	..	..	..	..	..	..
New Zealand[c]	1.4	..	..	..	..	..	..	12.3	..
Ireland	2.2	10	40.5	2.6	56.9	86.4	13.6	13.4	118.8
Iceland	0.0	0	..	..	..	..	..	..	..
Total OECD[a,b]	11 840	66 890	92	4½	3½	95½	4½	8	4

Source: OECD/STIIU data bank - December 1985.
a) OECD estimates.
b) See explanation table 11.
c) 1979.
d) Expenditure for 1980.
e) 1982.

Table 16. **Industry group: Chemical-linked - 1981**

	Million $	Researchers	Expenditure						
			Composition (%)			% Financed by		Compound growth rates	
			Food, drink and tobacco	Textiles and clothing	Rubber and plastics	Private	Public	1975-79	1979-81
United States[e]	1 391.0	17 400	44.6	8.3	47.1	88.9	11.1	1.4	..
Japan	1 115.5	19 857	38.2	24.7	37.1	99.4	0.6	12.3	15.0
Germany	339.1	2 148	36.2	15.3	48.5	91.3	8.1	20.5	9.1
France	333.3	1 680	22.4	12.2	65.5	98.2	1.4	3.4	2.3
United Kingdom	251.7	3 317	67.4	10.5	22.1	96.0	4.0	..	..
Italy	99.3	714	17.9	21.0	61.1	99.5	0.5	-3.3	-6.4
Canada	79.7	853	64.0	12.4	23.6	89.9	9.0	3.6	13.7
Spain	31.9	246	49.8	8.4	41.8	98.4	1.6	..	-24.1
Australia	19.8	300	66.8	4.1	29.1	87.8	..	..	..
Netherlands	95.8	606	86.2	3.9	10.0	..	..	2.6	-4.7
Turkey	..	..	..	..	..	..	..	..	..
Sweden	50.8	598	64.0	14.2	21.9	92.6	7.4	-0.9	2.7
Belgium	59.9	310	38.5	40.2	21.3	97.4	2.3	1.8	-2.2
Switzerland[d]	31.6	215	47.5	34.4	18.1	..	..	..	-30.3
Austria	39.9	251	25.4	15.0	59.6	94.3	5.7	..	..
Yugoslavia	..	..	..	..	..	..	..	..	..
Denmark	23.5	191	74.1	7.5	18.4	90.4	8.2	-1.2	9.6
Norway	14.2	133	67.2	14.0	18.8	82.9	16.2	-1.1	-5.6
Greece	1.0	..	47.5	45.7	6.8	100.0	0.0	..	..
Finland	21.4	224	57.2	13.8	29.0	94.6	5.1	7.3	7.8
Portugal[f]	1.5	20	..	..	..	..	..	..	..
New Zealand[c]	11.1	..	..	..	..	..	..	-6.1	..
Ireland	17.8	114	67.8	11.1	21.2	86.6	11.9	11.0	8.9
Iceland	0.1	3	53.8	15.4	30.8	53.8	46.2	..	5.4
Total OECD[a,b]	4 240	49 240	42	18½	39	97	2½	4½	5

Source: OECD/STIIU data bank - December 1985.
a) OECD estimates.
b) See explanation table 11.
c) 1979.
d) Researchers for 1979.
e) Expenditure for 1980.
f) 1982.

Table 17. **Industry group: Basic metals - 1981**

	Million $	Researchers	Expenditure						
			Composition (%)			% Financed by		Compound growth rates	
			Ferrous metals	Non-ferrous metals	Fabricated metal products	Private	Public	1975-79	1979-81
United States	1 502.0	16 600	35.5	22.7	41.5	83.0	17.0	2.1	7.5
Japan	1 290.8	12 389	56.2	22.4	21.4	97.8	2.1	5.2	15.6
Germany	484.3	2 680	38.6	12.0	49.4	73.7	24.7	18.5	8.8
France	207.6	1 179	34.7	31.4	33.8	91.9	4.0	-0.4	7.2
United Kingdom	168.5	2 403	34.8	23.7	41.4	92.8	7.2	..	..
Italy	64.9	459	46.4	53.6	0.0	93.1	2.6	-5.5	16.7
Canada	119.0	769	18.0	64.7	17.3	91.7	3.0	-1.2	7.3
Spain	30.7	150	..	..	..	93.0	0.0	..	-4.6
Australia	34.5	375	67.4	12.9	19.6	86.5	8.8	..	..
Netherlands	..	..	..	..	..	..	..	..	..
Turkey	..	..	..	..	..	..	..	..	..
Sweden	102.2	715	64.2	10.5	25.3	93.9	4.3	1.8	1.9
Belgium	76.6	301	42.1	18.4	39.5	92.8	7.2	3.4	-8.6
Switzerland[d]	58.2	300	83.8	5.4	10.8	..	..	..	-1.7
Austria	38.0	203	61.8	3.9	34.3	95.1	4.8	..	..
Yugoslavia	..	..	..	..	..	..	..	..	..
Denmark	5.4	26	..	..	..	93.9	6.1	6.4	14.4
Norway	35.0	268	30.8	50.0	19.2	82.7	12.6	6.9	1.6
Greece	2.8	..	11.1	7.4	81.5	100.0	0.0	..	..
Finland	22.5	219	30.4	39.7	29.9	91.3	8.6	8.2	13.2
Portugal[e]	1.2	27	..	..	..	..	..	..	..
New Zealand[c]	2.3	..	..	..	..	..	..	-9.8	..
Ireland	4.0	24	14.6	1.0	84.4	85.1	14.9	0.6	6.4
Iceland	0.5	9	4.8	9.5	85.7	66.7	33.3	-22.0	69.2
Total OECD[a,b]	4 250	39 020	52	15	33½	88	11	3½	8½

Source: OECD/STIIU data bank - December 1985.
a) OECD estimates.
b) See explanation table 11.
c) Expenditure for 1979.
d) Researchers for 1979.
e) 1982.

Table 18. **Industry group: Services - 1981**

	Million $	Researchers	Expenditure										
			Composition (%)						% Financed by		Compound growth rates	
			Utilities	Construction	Transport and storage	Communications	Commercial and engineering services	Other services	Private	Public	1975-79	1979-81
United States	1 906.0	22 200	..	..	..	..	..	..	57.0	43.0	12.5	1.8
Japan	1 016.8	8 809	22.1	30.6	7.5	39.1	..	0.6	97.2	2.5	5.8	5.6
Germany	253.2	2 630	22.2	16.4	17.3	..	38.9	5.3	54.1	44.3	1.4	−0.9
France	367.2	2 684	47.2	11.3	6.2	..	27.4	7.8	90.2	8.5	4.2	7.0
United Kingdom	335.3	4 166	65.3	8.5	9.0	4.4	0.0	12.9	94.7	5.3	..	..
Italy	422.3	3 697	22.7	6.7	0.2	4.4	65.8	0.1	83.5	13.2	8.1	4.9
Canada	258.7	2 848	31.8	..	26.3	..	25.6	16.3	83.4	14.2	6.9	18.6
Spain	62.3	579	7.7	10.6	3.4	4.3	58.1	15.7	82.8	15.3	..	7.1
Australia	110.0	1 029	..	25.7	..	..	..	74.3	85.7	..	..	..
Netherlands	100.4	830	3.9	16.7	..	..	..	16.0	..	..	6.7	6.7
Turkey	..	..	..	..	..	..	..	..	..	..	..	..
Sweden	159.4	1 088	49.0	..	..	..	36.9	3.4	60.4	30.1	−2.1	44.5
Belgium	115.6	425	0.2	4.6	0.2	0.0	41.1	53.9	94.0	5.5	48.5	5.7
Switzerland[d]	24.8	55	2.7	3.4	0.0	0.0	24.7	69.1	..	..	..	..
Austria	30.7	260	2.1	12.7	0.0	0.0	..	17.0	42.0	52.8	..	..
Yugoslavia	35.2	1 057	..	..	83.0	..	..	17.0	59.7	20.3	..	..
Denmark	53.4	601	..	3.8	..	9.1	8.4	78.8	58.0	34.1	9.2	4.8
Norway	34.3	459	19.5	26.0	16.3	18.7	15.2	4.3	31.6	66.2	4.4	−9.0
Greece	7.0	..	81.4	..	..	18.6	..	..	0.0	100.0	..	..
Finland	15.5	290	26.7	5.2	2.4	27.6	..	..	92.0	7.2	17.9	0.2
Portugal[e]	15.9	316	..	..	..	..	..	..	..	..	..	..
New Zealand[c]	6.0	..	..	..	..	..	..	..	..	..	..	..
Ireland	5.7	53	54.7	3.0	13.9	4.6	23.8	0.0	92.5	3.2	3.8	3.1
Iceland	0.0	0	..	..	..	..	..	..	..	..	..	..
Total OECD[a,b]	5 300	58 978	28½	15	8½	14	25	9½	84	14	7½	3½

Source: OECD/STIIU data bank - December 1985.
a) OECD estimates.
b) See explanation table 11.
c) 1979.
d) Researchers for 1979.
e) 1982.

Table 19. **Higher education R&D (HERD): national trends**
All fields of science

	1981		Expenditure growth rates (%)					1983
	Million $	Researchers	1969-81	1969-75	1975-81	1979-81	1981-83	Million $
United States	10 649.0	98 700	1.2	−0.9	3.4	3.6	0.0	11 863.0
Japan[b]	6 180.1	163 264	6.8	8.4	5.2	4.4	5.5	7 694.5
Germany	2 635.8	30 229	4.2	6.7	1.8	5.4	−1.3	2 867.1
France	1 756.8	32 700	2.7[a]	2.6	3.2[a]	3.6[a]	2.9[a]	2 081.0
United Kingdom[a]	1 512.5	..	5.2	7.5	2.8	4.5	1.2	1 733.8
Italy	814.2	24 754	1.8	2.5	1.2	12.7	..	..
Canada	978.3	7 520	−0.9	−2.7	1.0	−0.5	3.1	1 154.6
Spain	158.2	7 518	25.1	30.3	..	5.4	..	..
Australia	457.6	13 610	..	..	4.1	1.5	..	..
Netherlands	581.9	6 123	3.0	5.3	0.8	−3.3	7.3	746.2
Turkey	..	..	..	..	..	..	..	..
Sweden (n.s.e.)	580.8	5 400	8.1	12.6	3.8	8.5	7.5	748.8
Belgium[c]	220.0	6 538	..	−0.9	..	..	..	..
Switzerland[d]	355.0	4 860	6.0	6.5	5.5	6.5	−6.8	345.2
Austria	251.0	3 050	10.0	15.6	4.6	4.6	..	..
Yugoslavia	98.8	7 567	..	..	..	..	..	..
Denmark[c]	108.3	2 397	..	..	..	..	..	..
Norway	171.7	2 899	4.1	6.2	2.0	−4.1	0.8	197.0
Greece[d]	14.8	1 270	..	..	..	..	..	..
Finland[f]	111.0	3 733	7.5	6.7	2.3	2.6	5.0	136.5
Portugal[e]	31.8	1 267	0.5	−8.5	10.4	10.1	..	..
New Zealand[c]	37.9	1 130	..	..	..	..	..	..
Ireland	24.9	1 349	5.6	9.1	2.3	−1.1	..	..
Iceland	4.7	128	13.7	23.6	4.6	1.8	..	..
Total OECD[a]	27 740	427 720	3	3	3½	4½	2	..
of which EEC[a]	7 780	107 380	3½	5	2½	6	2	..

Source: OECD/STIIU data bank - December 1985.
a) OECD estimates.
b) Researchers not in full-time equivalent: OECD estimate is 81 632. Expenditure data adjusted by OECD would be 4 013 million $ in 1981, and 4 954 in 1983.
c) 1979
d) Researchers for 1979
e) 1982
f) Researchers for 1983

Table 20. **R&D resources in the higher education sector - 1983**
by major field of science

	Natural sciences and engineering [b]				Social sciences and humanities			
	Million $	Researchers	Percentage of total		Million $	Researchers	Percentage of total	
			Expenditure	Researchers			Expenditure	Researchers
United States	11 299	92 100	95	91	564	9 000	5	9
Japan	4 797	114 183	62	65	2 898	61 658	38	35
Germany [h]	2 127	20 762	81	69	508	9 537	19	31
France	..	..	..	..	..	..	..	..
United Kingdom [d,i]	1 167	12 400 [a]	67	72	578 [a]	4 800	33	28
Italy [h]	627	16 556	77	67	187	8 198	23	33
Canada	825	4 440	71	92	330	380	29	8
Spain [h]	107	4 660	67	62	52	2 858	33	38
Australia [h]	319	8 474	70	62	139	5 136	30	38
Netherlands [j,l]	541	3 957	73	65	197	2 166	27	35
Turkey	..	..	..	..	..	..	..	..
Sweden	749	5 800	85	81	129	1 400	15	19
Belgium [e]	184	5 106	84	78	37	1 432	17	22
Switzerland [g,i]	309	4 000	87	82	46	860	13	18
Austria [h]	193	2 065	77	68	58	986	23	32
Yugoslavia [h]	89	5 763	90	76	9	1 804	9	24
Denmark [k]	113	1 700	71	63	47	1 019	29	37
Norway [h]	128	1 932	74	67	43	967	25	33
Greece [g,i]	11	953	73	75	4	318	27	25
Finland	104	..	74	..	36	..	26	..
Portugal [k]	27	..	84	..	5	..	15	..
New Zealand [c,f]	19	611	66	54	10	519	34	46
Ireland [k]	21	895	78	65	7	488	26	35
Iceland [h]	4	94	81	73	1	34	19	27

Source: OECD/STIIU data bank - December 1985.
a) OECD estimates.
b) Natural sciences and engineering (n.s.e.) includes natural sciences, engineering, medical sciences and agricultural sciences.
c) Researchers for 1977.
d) Researchers for 1978.
e) 1979.
f) Expenditure for 1979.
g) Researchers for 1979.
h) 1981.
i) Expenditure for 1981.
j) Researchers for 1981.
k) 1982.
l) Expenditure for 1982.

Table 21

Estimated R&D expenditure in the higher education sector 1981 by main field of science

	A. By main OECD area (As a percentage of total OECD by field of science)				
	Natural sciences	Engineering	Medical sciences	Agricultural sciences	Social sciences and humanities
United States	54	37	31	52	14
Japan	4	19	18	12	34
EEC	30	32	36	21	35
Other countries	11	12	15	15	17
OECD	100	100	100	100	100

	B. Patterns in the main OECD area (As a percentage of total R&D in each area)					
	Natural sciences	Engineering	Medical sciences	Agricultural sciences	Social sciences and humanities	Total
United States	51	17	17	8	6	100
Japan	9	22	25	5	39	100
EEC	35	18	24	4	19	100
Other countries	31	16	23	7	23	100
OECD	37	18	22	6	17	100

Source: OECD/STIIU Data Bank, November 1985.

Table 22
Estimated R&D expenditure by type of activity in the OECD area 1981

	A. Share of each OECD area by type of activity (when total OECD by type of activity = 100)				B. Share of each OECD area in basic research by sector of performance (when total OECD by sector = 100)			
	Basic research	Basic + applied research	Experimental development		Business enterprise	Government	Higher education	Total
United States	44	45	56	United States	46	34	44	44
Japan	10	11	13	Japan	19	7	9	10
EEC	36	34	24	EEC	31	45	36	36
Other countries	10	10	7	Other countries	4	14	11	10
OECD	100	100	100	OECD	100	100	100	100

Source: OECD/STIIU Data Bank, November 1985.

Table 23. **National patent applications**

	1965	1970	1975	1980	1981	1982	1983
United States	94 629	103 175	101 014	106 218	108 673	112 234	106 314
Japan	81 923	130 829	159 821	193 779	219 877	238 880	256 528
Germany	66 470	66 132	60 095	66 765	66 926	71 262	73 334
France	47 793	47 283	40 437	45 081	47 190	47 496	49 330
United Kingdom	55 507	62 101	53 400	59 643	62 356	62 721	63 241
Italy	29 308	31 828	24 151	29 943	32 007	31 961	32 894
Canada	30 093	30 510	25 652	24 974	25 498	25 293	25 707
Spain	13 630	11 810	10 522	10 877	10 227	10 201	9 146
Australia	15 150	16 443	14 082	14 781	18 092	18 084	18 368
Netherlands	17 284	19 109	15 267	21 263	23 790	24 339	25 887
Turkey	739	636	610	661	525	511	..
Sweden	17 079	17 858	14 799	21 334	23 159	23 715	25 477
Belgium	16 810	17 187	13 224	16 621	18 046	20 085	20 994
Switzerland	18 180	19 406	16 940	21 086	22 842	23 281	24 377
Austria	11 832	11 786	9 911	15 693	17 925	17 959	19 584
Yugoslavia	2 193	3 214	3 448	3 358	3 156	2 984	2 532
Denmark	6 713	6 637	5 958	6 590	7 323	7 190	7 539
Norway	4 899	5 007	4 431	4 738	5 724	5 733	6 307
Greece	1 964	2 672	2 981	2 898	3 154	3 260	3 211
Finland	3 145	3 528	3 761	4 218	5 099	5 651	6 067
Portugal	1 176	1 995	1 476	1 823	1 933	1 826	1 851
New Zealand	3 450	3 573	3 322	3 438	3 457	3 533	3 739
Ireland	1 363	1 662	2 844	2 749	3 110	3 110	3 094
Iceland	..	78	48	77	93	80	94

Source: OECD/STIIU data bank - December 1985.

Table 24. **Domestic patent applications**

	1965	1970	1975	1980	1981	1982	1983
United States	72 317	76 195	64 445	62 098	62 404	63 316	59 391
Japan	60 796	100 511	135 118	165 730	191 621	210 897	227 708
Germany	38 148	32 772	30 198	30 582	30 251	31 111	32 140
France	17 509	14 106	12 110	11 086	11 057	10 821	11 288
United Kingdom	24 274	25 227	20 842	19 710	20 898	20 640	20 011
Italy	7 473	7 241	5 977	6 375	..	..	..
Canada	1 854	1 986	1 853	1 785	1 951	1 936	2 017
Spain	4 089	2 966	1 903	1 876	1 718	1 646	1 369
Australia	4 123	3 984	4 311	6 582	6 341	6 603	6 930
Netherlands	2 505	2 462	1 966	1 995	2 073	2 093	2 118
Turkey	99	89	98	134	157	126	..
Sweden	4 814	4 343	4 042	4 126	3 954	4 119	4 331
Belgium	1 766	1 339	1 060	992	919	1 022	925
Switzerland	5 721	5 927	5 834	4 313	4 048	4 217	4 212
Austria	2 714	2 267	2 525	2 345	2 419	2 266	2 388
Yugoslavia	759	917	1 251	1 384	1 422	1 428	1 445
Denmark	1 153	815	828	964	1 085	1 095	1 167
Norway	870	938	752	716	714	693	825
Greece	888	1 411	1 664	1 308	1 273	1 291	1 251
Finland	819	861	1 164	1 356	1 423	1 638	1 719
Portugal	128	178	72	92	90	92	91
New Zealand	788	897	1 243	1 148	1 033	986	1 110
Ireland	157	205	329	394	461	434	567
Iceland	..	19	14	19	14	18	32

Source: OECD/STIIU data bank - December 1985.

Table 25. **Foreign patent applications**

	1965	1970	1975	1980	1981	1982	1983
United States	22 312	26 980	36 569	44 120	46 269	48 918	46 923
Japan	21 127	30 318	24 703	28 049	28 256	27 983	28 820
Germany	28 322	33 360	29 897	36 183	36 675	40 151	41 194
France	30 284	33 177	28 327	33 995	36 133	36 675	38 042
United Kingdom	31 233	36 874	32 558	39 933	41 458	42 081	43 230
Italy	21 835	24 587	18 174	23 568	..	..	..
Canada	28 239	28 524	23 799	23 189	23 547	23 357	23 690
Spain	9 541	8 844	8 619	9 001	8 509	8 555	7 777
Australia	11 027	12 459	9 771	10 205	11 751	11 481	11 438
Netherlands	14 779	16 647	13 301	19 268	21 717	22 246	23 769
Turkey	640	547	512	527	368	385	..
Sweden	12 265	13 515	10 757	17 208	19 205	19 596	21 146
Belgium	15 044	15 848	12 164	15 629	17 127	19 063	20 069
Switzerland	12 459	13 479	11 106	16 773	18 794	19 064	20 165
Austria	9 118	9 519	7 386	13 348	15 506	15 693	17 196
Yugoslavia	1 434	2 297	2 197	1 974	1 734	1 556	1 087
Denmark	5 560	5 822	5 130	5 626	6 238	6 095	6 372
Norway	4 029	4 069	3 679	4 022	5 010	5 040	5 482
Greece	1 076	1 261	1 317	1 590	1 881	1 969	1 960
Finland	2 326	2 667	2 597	2 862	3 676	4 013	4 348
Portugal	1 048	1 817	1 404	1 731	1 843	1 734	1 760
New Zealand	2 662	2 676	2 089	2 290	2 424	2 547	2 683
Ireland	1 206	1 457	2 505	2 355	2 649	2 676	2 757
Iceland	..	59	34	58	79	62	62

Source: OECD/STIIU data bank - December 1985.

Table 26. **External patent applications**

	1965	1970	1975	1980	1981	1982	1983
United States	103 214	123 724	93 042	116 337	126 990	123 241	135 532
Japan	8 421	26 568	27 666	45 465	49 315	56 411	55 312
Germany	52 933	70 137	60 810	82 594	82 601	79 530	76 700
France	19 631	24 422	23 433	32 956	31 386	34 700	34 346
United Kingdom	32 311	33 463	24 402	28 134	31 230	33 221	33 648
Italy	7 322	10 477	10 080	12 569	13 373	14 147	13 537
Canada	4 672	4 830	5 063	4 577	5 137	5 387	5 628
Spain	936	1 406	1 759	1 599	1 848	1 477	1 540
Australia	1 325	1 933	2 131	3 698	4 515	4 888	5 039
Netherlands	11 321	12 059	9 908	12 150	12 146	12 030	12 703
Turkey	..	..	..	..	..	30	10
Sweden	7 505	8 508	9 328	11 377	12 399	12 745	13 508
Belgium	3 843	3 872	3 197	4 062	3 677	3 764	3 871
Switzerland	19 180	25 910	19 729	22 820	21 259	21 138	21 689
Austria	2 677	3 970	3 277	4 551	4 610	4 182	4 779
Yugoslavia	224	..	189	268	260	278	179
Denmark	2 113	2 613	2 297	2 878	3 244	3 677	3 805
Norway	854	1 191	1 252	1 287	1 494	1 646	1 500
Greece	70	..	168	138	137	180	119
Finland	571	1 223	1 345	2 122	2 562	3 059	3 216
Portugal	96	160	656	10	5	4	10
New Zealand	286	446	490	757	650	672	808
Ireland	151	260	168	497	527	394	426
Iceland	..	..	..	..	7	7	7

Source: OCDE/STIIU data bank - December 1985.

Table 27. **Number of patents granted per 100 patent applications**

	1965	1970	1975	1980	1981	1982
United States	66	62	71	59	62	53
Japan	33	24	29	24	24	22
Germany	25	19	30	39	29	34
France	87	56	35	100	87	108
United Kingdom	61	66	76	57	58	80
Italy	69	93	..	49	44	44
Canada	81	96	80	96	89	87
Spain	79	62	88	85	71	95
Australia	48	37	86	53	40	35
Netherlands	14	13	25	45	50	131
Turkey	81	67	76	73	53	68
Sweden	46	77	61	54	62	104
Belgium	99	99	99	99	99	99
Switzerland	104	91	81	62	102	129
Austria	59	74	71	92	94	84
Yugoslavia	30	24	14	20	26	9
Denmark	41	51	41	30	25	26
Norway	49	49	50	53	41	42
Greece	105	108	61	71	80	69
Finland	26	39	36	47	48	45
Portugal	83	102	206	126	82	66
New Zealand	60	95	69	37	47	39
Ireland	56	47	34	52	43	35
Iceland	..	97	44	35	31	28
OECD average	57	53	50	49	46	46

Source: OCDE/STIIU data bank -December 1985.
Note: WIPO-data only.

Table 28. **Domestic patent applications per 100 000 population**

	1965	1970	1975	1980	1981	1982	1983
United States	37	37	30	27	27	27	25
Japan	62	97	121	142	163	178	191
Germany	65	54	49	50	49	50	52
France	36	28	23	21	20	20	21
United Kingdom	45	45	37	35	37	37	35
Italy	14	13	11	11	..	..	..
Canada	9	9	8	7	8	8	8
Spain	13	9	5	5	5	4	4
Australia	35	31	31	45	42	44	45
Netherlands	20	19	14	14	15	15	15
Turkey	0	0	0	0	0	0	..
Sweden	62	54	49	50	48	49	52
Belgium	19	14	11	10	9	10	9
Switzerland	96	95	91	68	63	65	65
Austria	37	30	33	31	32	30	32
Yugoslavia	4	5	6	6	6	6	6
Denmark	24	17	16	19	21	21	23
Norway	23	24	19	18	17	17	20
Greece	10	16	18	14	13	13	13
Finland	18	19	25	28	30	34	35
Portugal	1	2	1	1	1	1	1
New Zealand	30	32	40	37	33	31	34
Ireland	5	7	10	12	13	12	16
Iceland	..	9	6	8	6	8	14
OECD average	36	39	39	41	43	45	47

Source: OECD/STIIU data bank -December 1985.

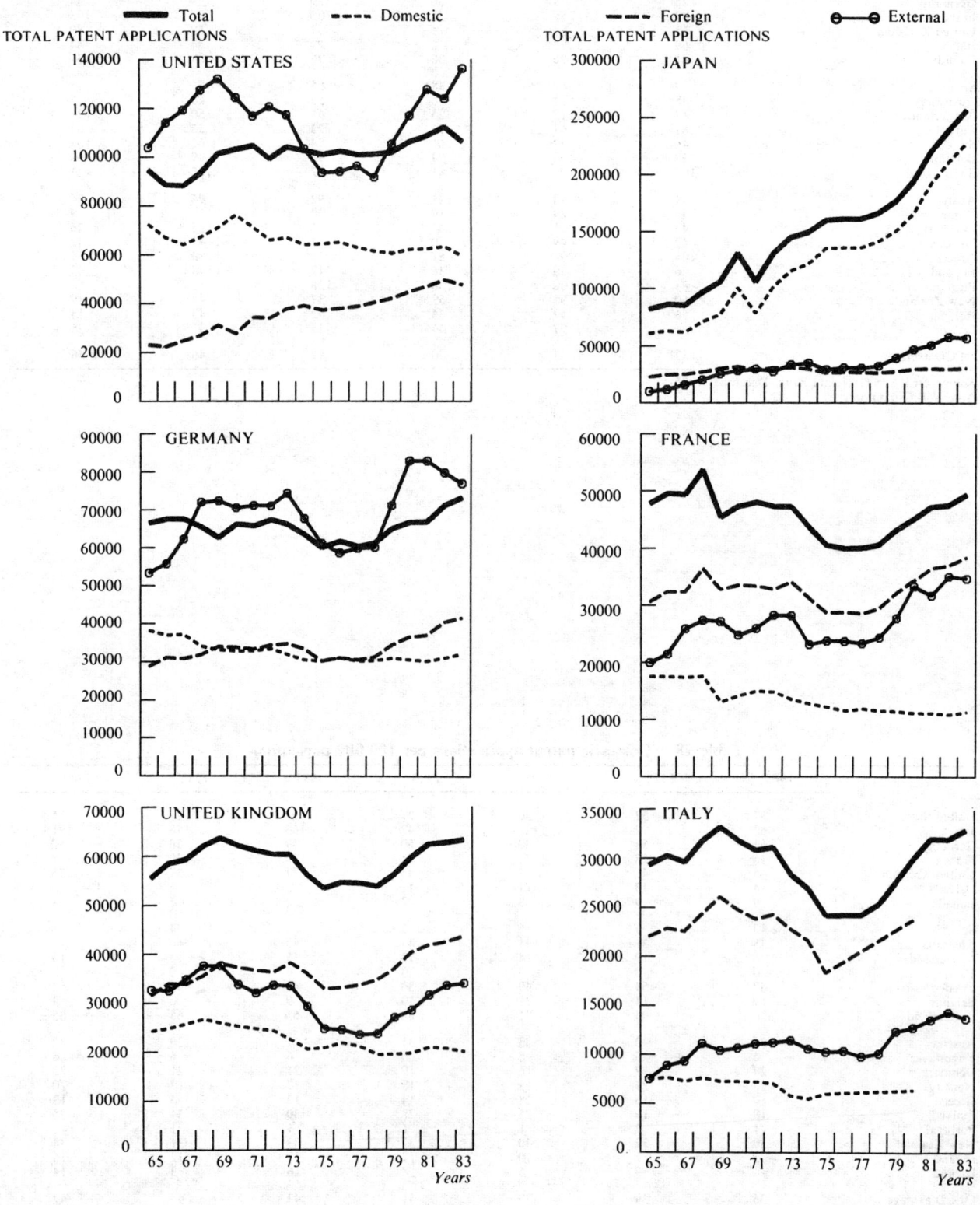

Graph 1
Patents applications by type in OECD Member countries 1965 to 1983
Including from 1978 onwards international patent channels

Source: OECD/STIIU Data Bank, November 1985.

Graph 1 (cont'd)

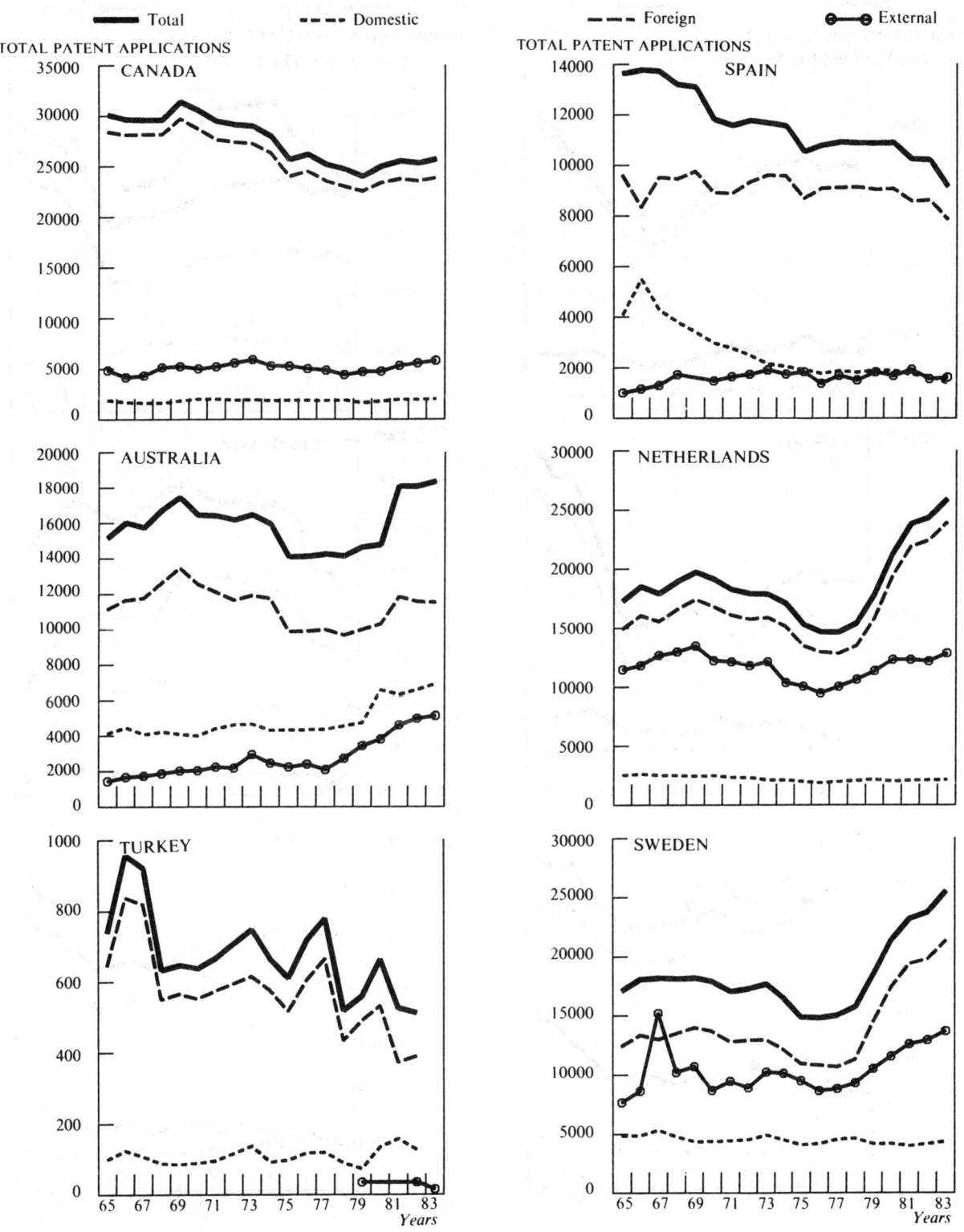

Graph 1 (cont'd)

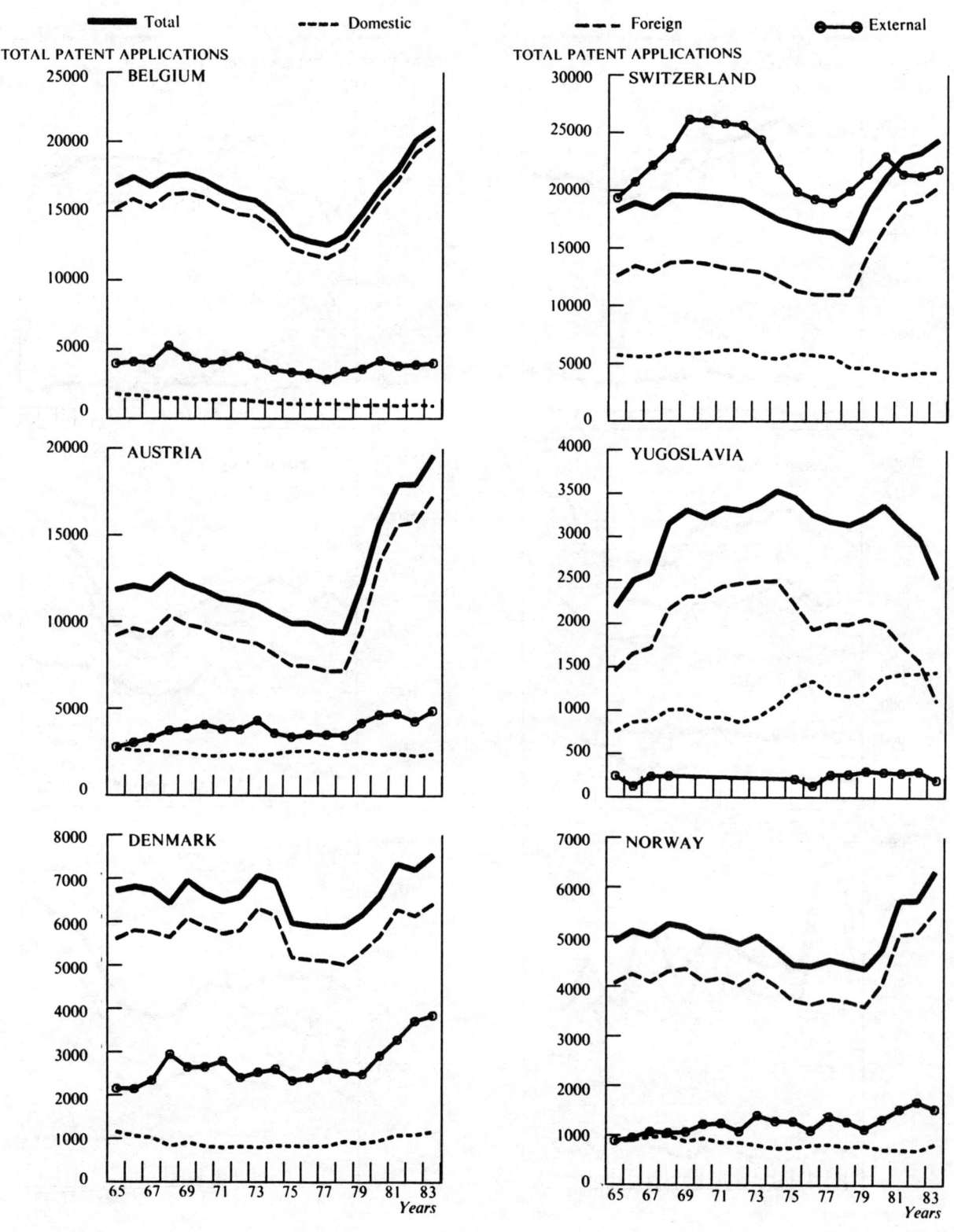

Graph 1 (cont'd)

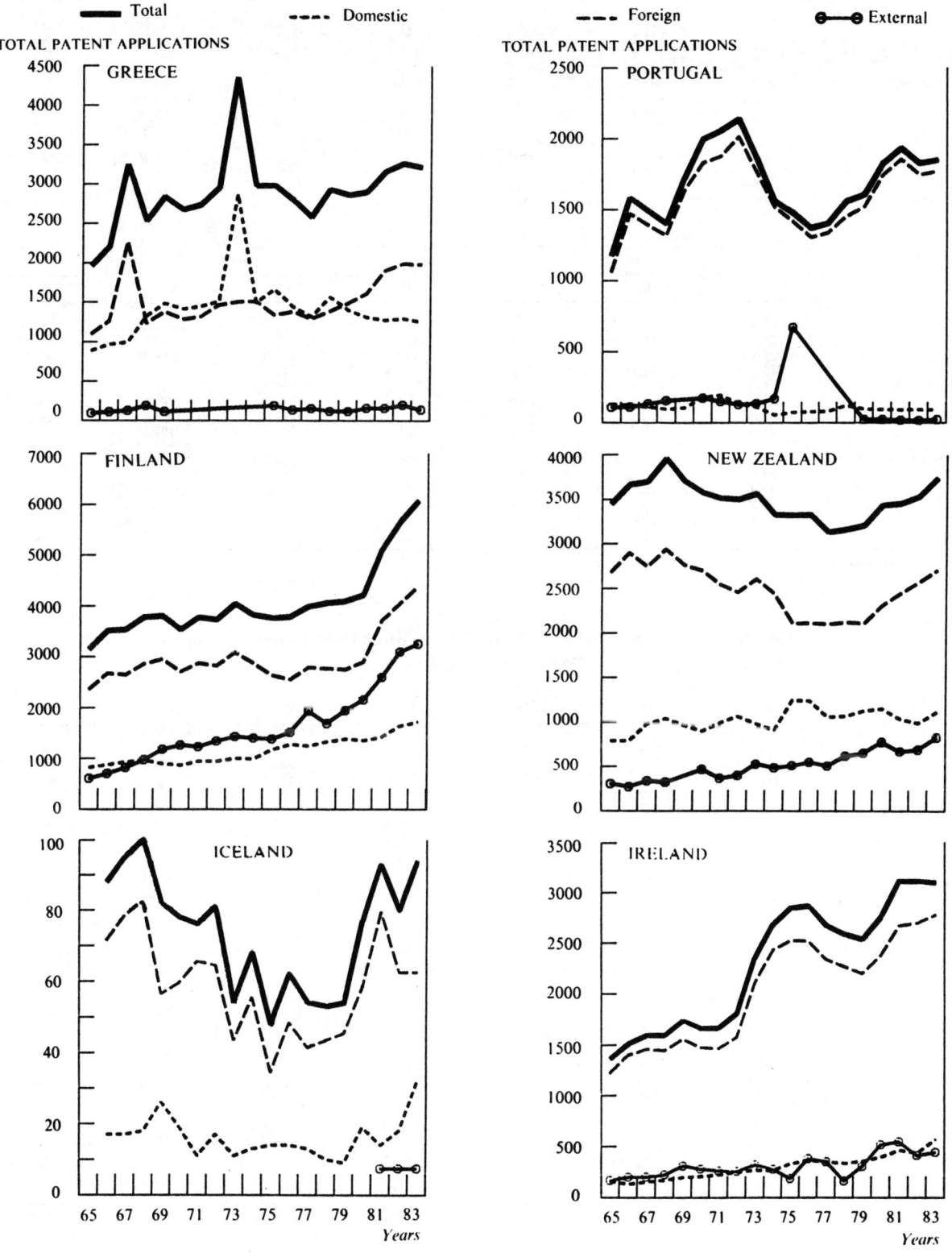

Table 29. **Technological balance of payments of fifteen OECD member countries**
Millions of current dollars
Receipts

	1970	1971	1972	1973	1974	1975	1976
United States	2 134.0	2 375.0	2 566.0	3 021.0	3 584.0	4 008.0	4 084.0
Japan	..	111.1	170.9	194.3	197.3	232.9	290.3
Germany	152.5	174.7	213.9	187.3	229.4	263.5	259.3
France	172.8	175.7	240.1	275.9	392.0	393.6	487.1
United Kingdom	406.8	404.3	447.6	545.0	614.0	588.6	813.3
Italy	..	..	64.4	77.9	101.5	90.2	113.4
Canada	..	34.1	..	38.1	..	53.1	..
Spain	..	..	..	42.8	49.6	64.5	82.3
Australia	..	..	..	..	..	..	7.6
Netherlands	140.0	139.7	119.7	139.0	150.6	159.7	184.4
Sweden	..	19.2	..	20.2	..	23.2	..
Austria	9.8	9.7	9.1	9.0	12.1	10.0	17.1
Denmark	35.0	39.2	38.7	40.9	46.5	49.2	58.6
Finland	2.4	2.6	2.5	3.1	3.2	3.0	12.3
Portugal	..	..	1.5	2.3	2.6	..	..

	1977	1978	1979	1980	1981	1982	1983
United States	4 503.0	5 312.0	5 747.0	6 617.0	6 863.0	6 878.0	7 531.0
Japan	325.1	436.6	503.6	643.6	748.6	830.9	1 123.6
Germany	286.1	328.7	357.3	430.2	503.0	566.4	635.8
France	575.3	636.7	657.9	687.5	843.1	913.5	969.0
United Kingdom	821.7	849.6	791.2	778.9	890.3	928.9	1 131.0
Italy	201.9	163.5	194.6	233.4	252.8	220.8	207.7
Canada	62.6	..	86.8	..	139.6	209.0	210.9
Spain	78.2	86.7	110.8	151.6	223.2	197.5	214.8
Australia	..	12.0	..	..	12.3	..	..
Netherlands	188.1	212.6	261.9	304.0	364.4	347.5	..
Sweden	27.1	..	23.7	..	58.2	..	101.0
Austria	19.9	20.3	23.1	23.7	23.6	33.2	29.5
Denmark	65.0	66.5	..	..	93.7	..	..
Finland	16.1	24.6	37.7	42.3	46.2	47.1	..
Portugal	4.1	2.9	4.8	5.5	8.2	7.3	8.4

Source: OECD/STIIU data bank -December 1985.

Table 30. **Technological balance of payments of fifteen OECD member countries**
Millions of current dollars
Payments

	1970	1971	1972	1973	1974	1975	1976
United States	225.0	241.0	294.0	385.0	346.0	473.0	482.0
Japan	..	549.6	704.9	870.2	552.1	591.6	617.2
Germany	412.7	475.3	499.5	519.8	558.4	679.2	687.7
France	282.2	324.3	365.4	400.4	477.0	471.7	616.0
United Kingdom	379.7	378.5	405.6	465.5	546.0	578.3	645.9
Italy	..	..	374.5	394.4	370.8	479.7	456.5
Canada	..	121.1	..	158.5	..	189.8	..
Spain	..	..	..	388.0	432.9	386.7	633.0
Australia	..	..	..	..	..	..	93.4
Netherlands	162.2	154.1	176.5	186.2	221.2	243.5	313.9
Sweden	..	17.1	..	18.5	..	26.6	..
Austria	51.2	41.9	48.3	49.1	63.8	71.7	77.5
Denmark	28.9	29.8	35.3	32.9	34.3	36.8	45.9
Finland	16.2	17.8	18.6	28.2	29.6	35.4	40.1
Portugal	..	..	18.2	21.4	28.4	..	..

	1977	1978	1979	1980	1981	1982	1983
United States	434.0	610.0	764.0	762.0	693.0	200.0	230.0
Japan	662.1	687.1	911.5	965.8	1 109.9	1 269.9	1 302.7
Germany	792.6	850.9	935.8	1 025.3	1 087.6	1 124.8	1 269.1
France	629.2	686.1	767.7	822.7	922.3	1 065.4	1 073.5
United Kingdom	672.7	713.9	658.2	672.1	736.1	767.6	887.2
Italy	576.8	826.8	597.3	661.9	726.4	822.6	842.1
Canada	240.0	..	324.2	..	443.0	447.6	449.4
Spain	501.6	475.3	503.1	618.8	700.1	993.1	1 038.7
Australia	..	133.4	..	..	125.5	..	..
Netherlands	292.2	325.0	392.5	466.4	558.9	583.8	..
Sweden	28.0	..	26.7	..	54.6	..	46.2
Austria	87.1	88.7	96.7	109.1	98.1	115.7	154.2
Denmark	53.4	52.2	..	..	62.5	..	..
Finland	39.4	42.8	48.6	59.6	72.0	73.5	..
Portugal	30.1	27.6	35.1	45.4	57.7	74.4	83.4

Source: OECD/STIIU data bank -December 1985.

Table 31. **Technological balance of payments of fifteen OECD member countries**
Millions of current dollars
Balance

	1970	1971	1972	1973	1974	1975	1976
United States	1 909.0	2 134.0	2 272.0	2 636.0	3 238.0	3 535.0	3 602.0
Japan	0.0	−438.6	−534.0	−675.9	−354.9	−358.7	−326.9
Germany	−260.2	−300.6	−285.6	−332.5	−329.1	−415.7	−428.4
France	−109.4	−148.6	−125.3	−124.6	−84.9	−78.1	−128.9
United Kingdom	27.1	25.7	42.0	79.6	68.0	10.3	167.4
Italy	0.0	0.0	−310.2	−316.5	−269.4	−389.5	−343.2
Canada	0.0	−87.0	0.0	−120.5	0.0	−136.7	0.0
Spain	0.0	0.0	0.0	−345.2	−383.3	−322.2	−550.6
Australia	0.0	0.0	0.0	0.0	0.0	0.0	−85.8
Netherlands	−22.2	−14.3	−56.8	−47.3	−70.6	−83.8	−129.6
Sweden	0.0	2.2	0.0	1.7	0.0	−3.4	0.0
Austria	−41.4	−32.2	−39.2	−40.0	−51.7	−61.7	−60.4
Denmark	6.1	9.4	3.4	8.0	12.2	12.4	12.7
Finland	−13.8	−15.2	−16.1	−25.1	−26.4	−32.4	−27.8
Portugal	0.0	0.0	−16.8	−19.2	−25.8	0.0	0.0

	1977	1978	1979	1980	1981	1982	1983
United States	4 069.0	4 702.0	4 983.0	5 855.0	6 170.0	6 678.0	7 301.0
Japan	−337.0	−250.5	−407.9	−322.2	−361.3	−439.0	−179.1
Germany	−506.5	−522.2	−578.5	−595.1	−584.6	−558.5	−633.3
France	−53.9	−49.4	−109.8	−135.2	−79.2	−151.9	−104.5
United Kingdom	149.1	135.7	133.0	106.8	154.2	161.3	243.9
Italy	−374.8	−663.2	−402.8	−428.6	−473.6	−601.7	−634.4
Canada	−177.4	0.0	−237.4	0.0	−303.4	−238.5	−238.5
Spain	−423.3	−388.6	−392.3	−467.3	−476.9	−795.6	−823.9
Australia	0.0	−121.3	0.0	0.0	−113.2	0.0	0.0
Netherlands	−104.1	−112.4	−130.6	−162.4	−194.5	−236.3	0.0
Sweden	−1.0	0.0	−3.0	0.0	3.6	0.0	54.8
Austria	−67.2	−68.4	−73.6	−85.4	−74.5	−82.6	−124.8
Denmark	11.6	14.3	0.0	0.0	31.2	0.0	0.0
Finland	−23.3	−18.2	−11.0	−17.3	−25.8	−26.4	0.0
Portugal	−26.0	−24.8	−30.4	−39.9	−49.5	−67.1	−75.0

Source: OECD/STIIU data bank -December 1985.

Table 32. **Shares of related firms in total receipts and payments**
Percentages

	Receipts					Payments				
	1967	1970	1975	1980	1983	1967	1970	1975	1980	1983
United States	74.1	73.1	81.1	82.1	..	37.3	49.3	60.7	67.5	..
United Kingdom	35.3	33.2	31.8	39.1	47.2	58.1	59.1	65.1	84.9	84.8
Germany[a]	2.5	3.9	5.4	8.8	22.8	75.5	67.1	77.1	77.9	82.4

Source: OECD/STIIU data bank -December 1985.
a) In contrast to the United States and United Kingdom data, the share of German receipts is that of German firms with foreign participation in capital. Only payments are really comparable for all three countries above.

Pattern of trade
OECD are

Criteria	Weight in output			Weight in exports			Weight in imports		
Sub-classes	Low	Medium	High	Low	Medium	High	Low	Medium	H
Codes	YYL	YYM	YYH	XXL	XXM	XXH	MML	MMM	M
Values	0-3	3-8	8-26	0-3	3-10	10-20	0-3	3-10	10
High intensity									
Aerospace	×			×			×		
Computers	×			×			×		
Electronics – components		×			×			×	
Drugs and medicine	×			×			×		
Instruments	×			×			×		
Electrical machinery		×			×		×		
Medium intensity									
Motor vehicles		×				×			×
Chemicals		×				×		×	
Other manufacturing industries	×			×			×		
Non electrical machinery		×				×		×	
Rubber, plastics	×			×			×		
Non ferrous metals	×			×				×	
Low intensity									
Stone, clay, glass	×			×				×	
Food, drink			×		×				×
Shipbuilding	×			×			×		
Petroleum refineries		×			×			×	
Ferrous metals		×			×			×	
Fabricated metal products		×			×		×		
Paper, printing		×			×			×	
Wood, cork, furniture		×			×		×		
Textiles, footwear, leather		×			×			×	

Source: OECD/STIIU, November 1985.

nufacturing industries
countries

ight of exports in output			Weighty of imports in output			Balance in relation to trading average			Surplus output in relation to domestic demand			Import penetration rate			Export/import ratio		
ow	Medium	High	Low	Medium	High	Low	Medium	High	Low	Medium	High	Low	Medium	High	Low	Medium	High
L	XYM	XYH	MYL	MYM	MYH	BTL	BTM	BTH	YDL	YDM	YDH	PEL	PEM	PEH	TCL	TCM	TCH
10	11-35	36-69	15-22	23-30	30-81	-39 to 6	6-35	35-105	0-1	1-1.07	1.07-1.2	7-13	13-22	22-36	0-110	1-115	115-215
		x			x		x				x		x		x		
	x				x	x				x				x	x		
		x			x	x				x				x	x		
x			x			x				x			x				x
	x				x	x					x			x	x		
x			x					x			x	x					x
		x		x				x			x			x			x
		x			x			x			x		x			x	
x				x		x			x					x		x	
	x			x				x			x		x				x
x			x			x			x			x					x
		x		x		x					x	x				x	
x				x		x			x			x			x		
x				x				x			x			x			x
x			x			x			x			x			x		
x			x					x		x			x				x
x			x					x		x		x					x
x			x			x			x			x			x		
x				x		x			x			x			x		
x					x	x			x					x	x		

99

Graph 2
R&D intensities and weights of high R&D intensity industries in total exports of manufacturing industry

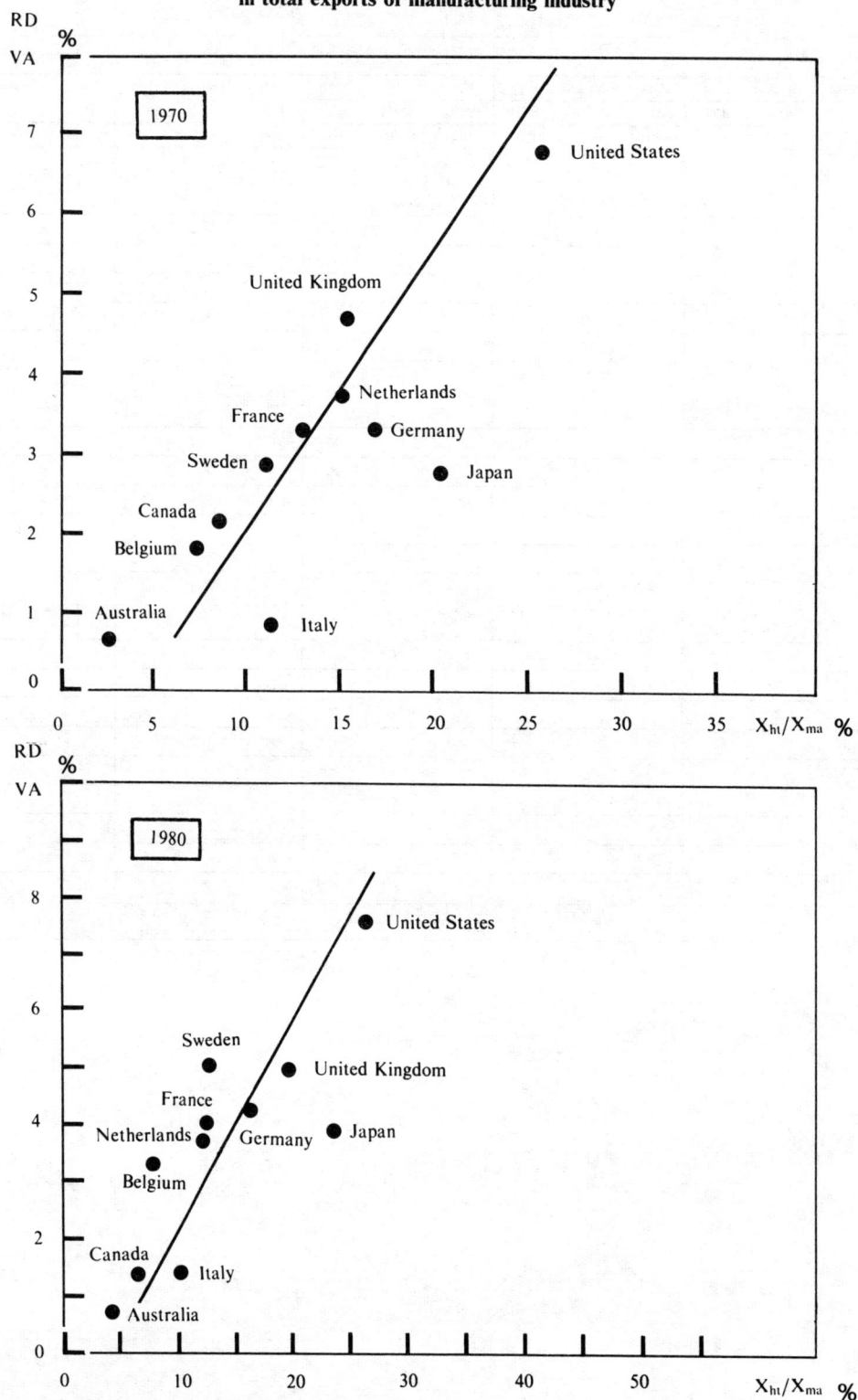

Where: VA: Value added of manufacturing industries.
RD: Total R&D expenditure of manufacturing industries.
Xht: Exports by high-intensity industries.
Xma: Exports by manufacturing industries.
Source: OECD/STIIU Data Bank, November 1985.

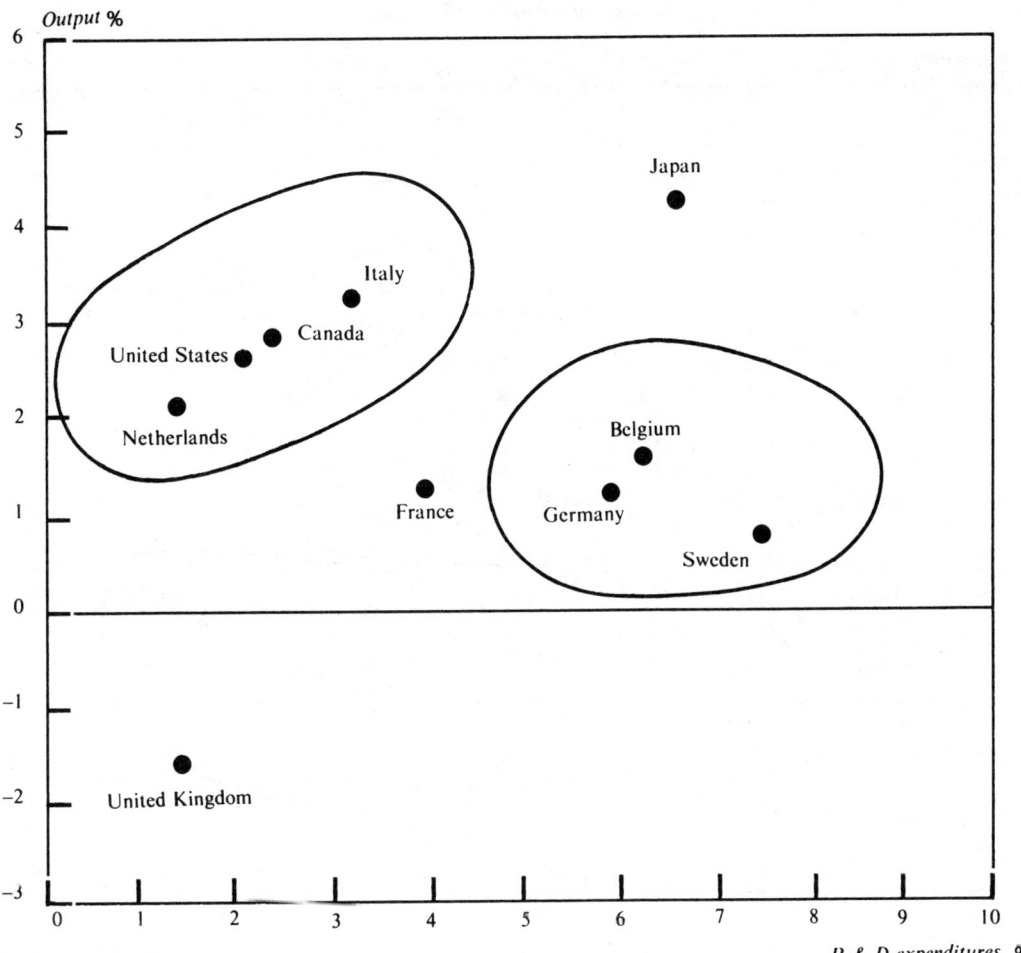

Graph 3
Growth in manufacturing output and R&D expenditure
Average annual rates 1972-1981*

* Or closest approximations.
Source: OECD/STIIU Data Bank, November 1985.

Table 34

Employment weights of high, medium and low R&D intensity industries in manufacturing industry

	High intensity			Medium intensity			Low intensity		
	1970	1975	1982	1970	1975	1982	1970	1975	1982
United States	18.4	18.4	21.6c	32.0	31.0	32.9c	49.2	50.5	45.5c
Japan	16.7	16.0	18.2b	33.3	32.0	33.3b	50.1	52.0	48.5b
Germany	16.3	16.6	17.5	33.0	34.2	39.8	50.7	49.2	42.7
France d	..	13.6a	14.1a	..	35.2	36.2	..	51.2	49.7
United Kingdom	16.8	17.2	19.9b	30.8	31.8	33.9b	52.4	51.0	46.2b
Italy c	..	13.7a	13.5c	..	30.0	32.9c	..	56.3	53.6c
Canada	13.0	11.9	12.9c	24.1	24.5	24.7c	62.9	63.6	62.4c
Australia	..	12.8	12.9c	..	22.6	22.6c	..	64.6	64.5c
Sweden	13.8	14.5	15.7c	28.6	30.4	31.3c	57.6	55.8	53.0c

a) Not including the aerospace industry.
b) 1980.
c) 1981.
d) Secretariat estimate.
Source: OECD, STIIU, November 1985.

Graph 4
Trends in productivity and numbers of RSE in manufacturing industry
Average annual rates, 1972-1981*

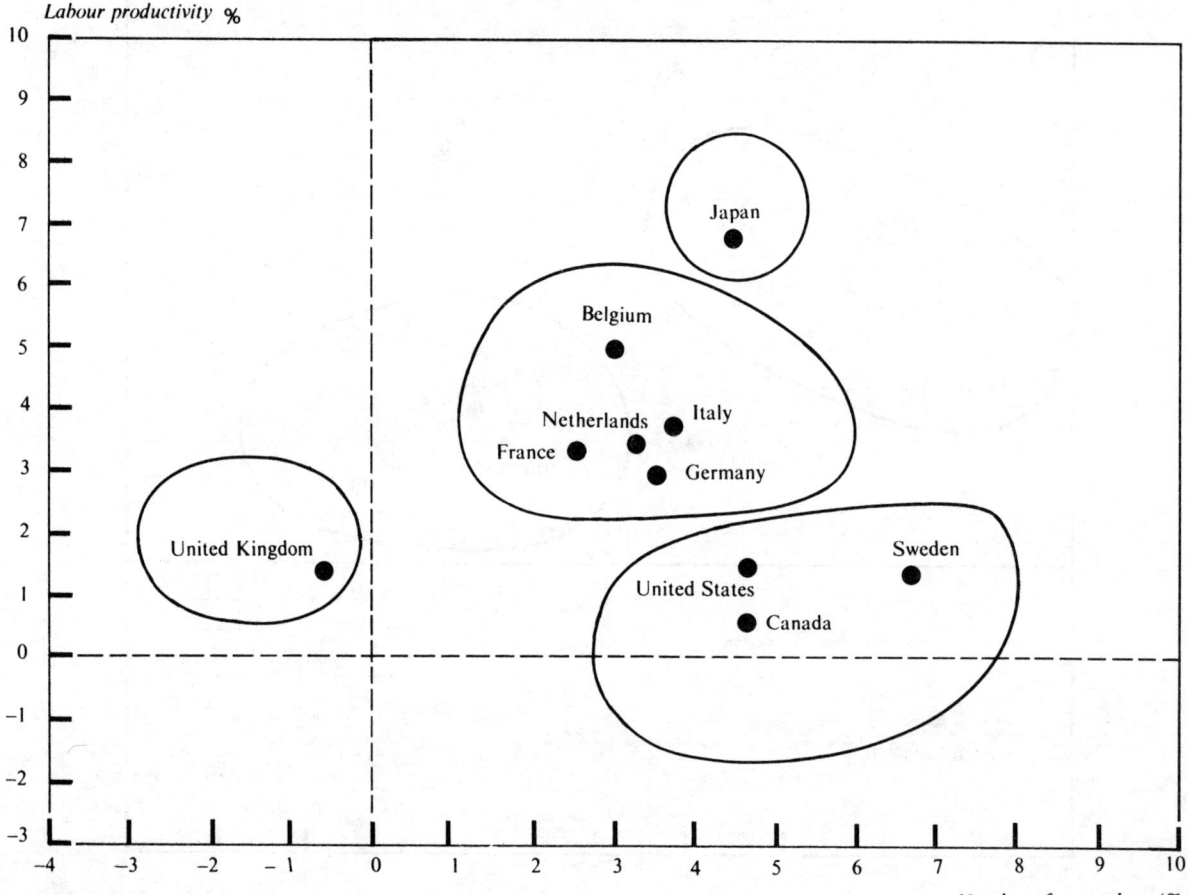

* Or closest approximations.
Source: OECD/STIIU Data Bank, November 1985.

Graph 5
**Labour productivity and R&D personnel intensity in
manufacturing industry, in 1970 and 1981**

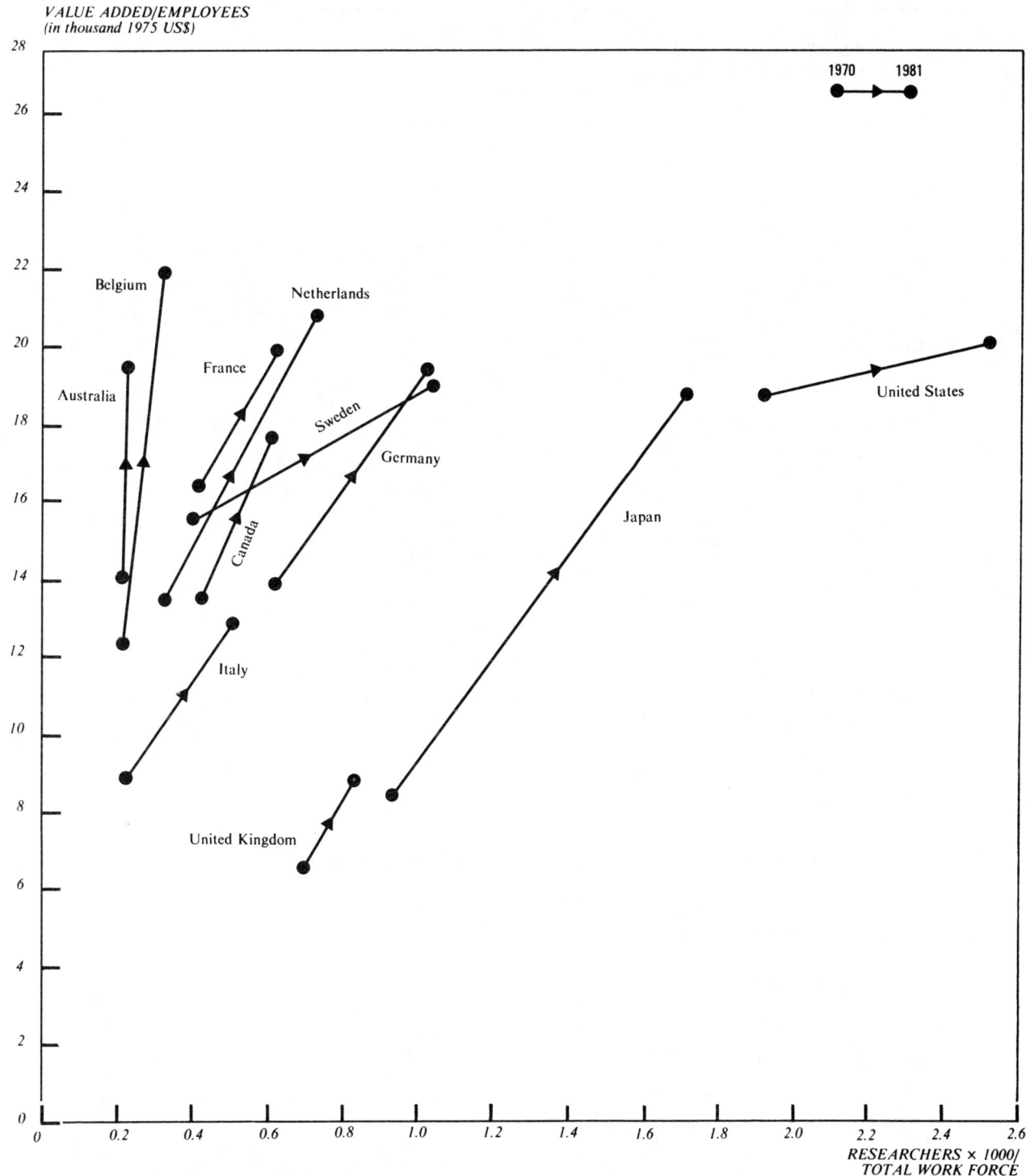

Source: OECD/STIIU Data Bank, November 1985.

Graph 6
Trade balance of manufacturing industries

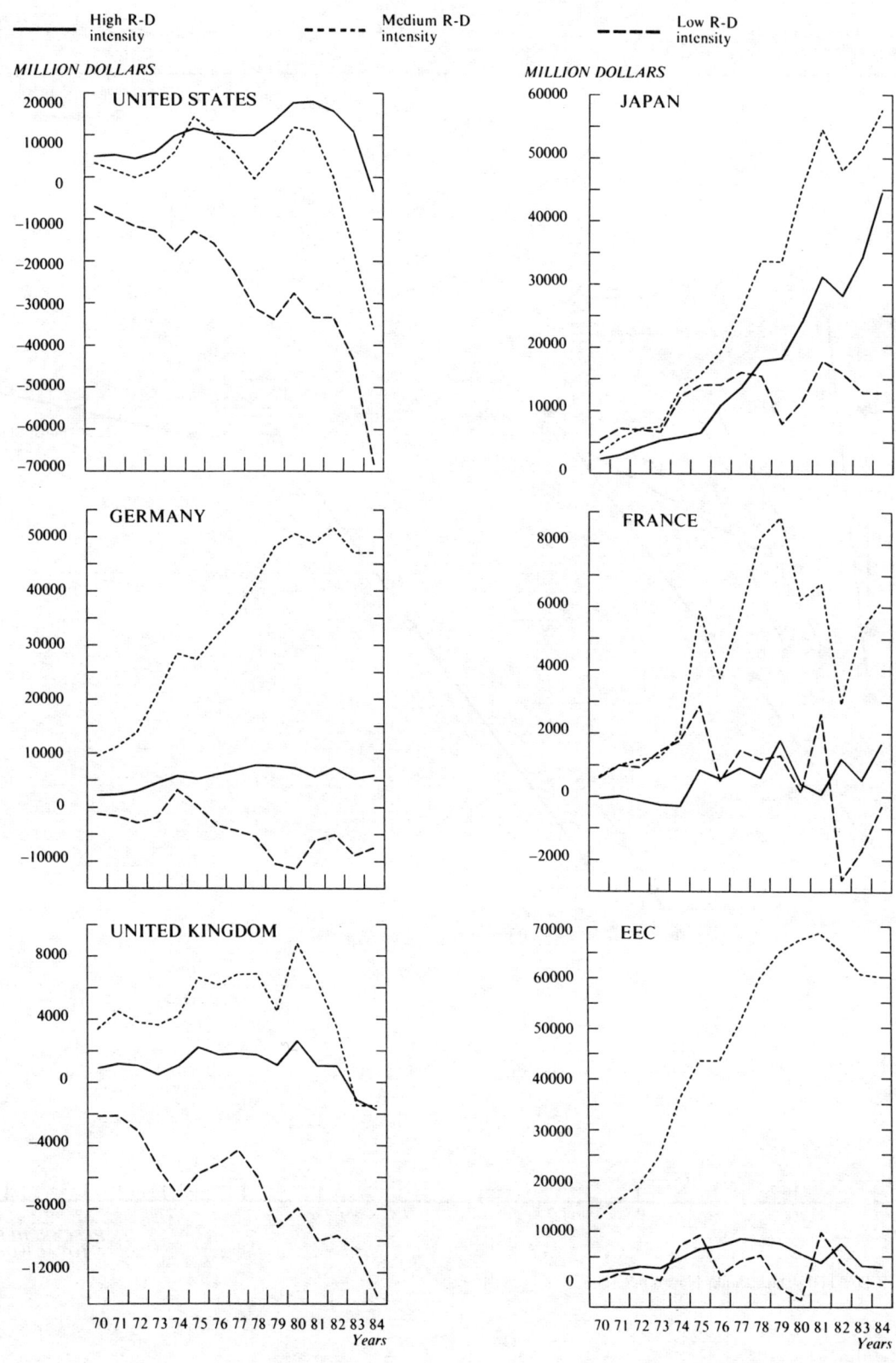

Source: OECD/STIIU, November 1985.

Graph 6 (cont'd)

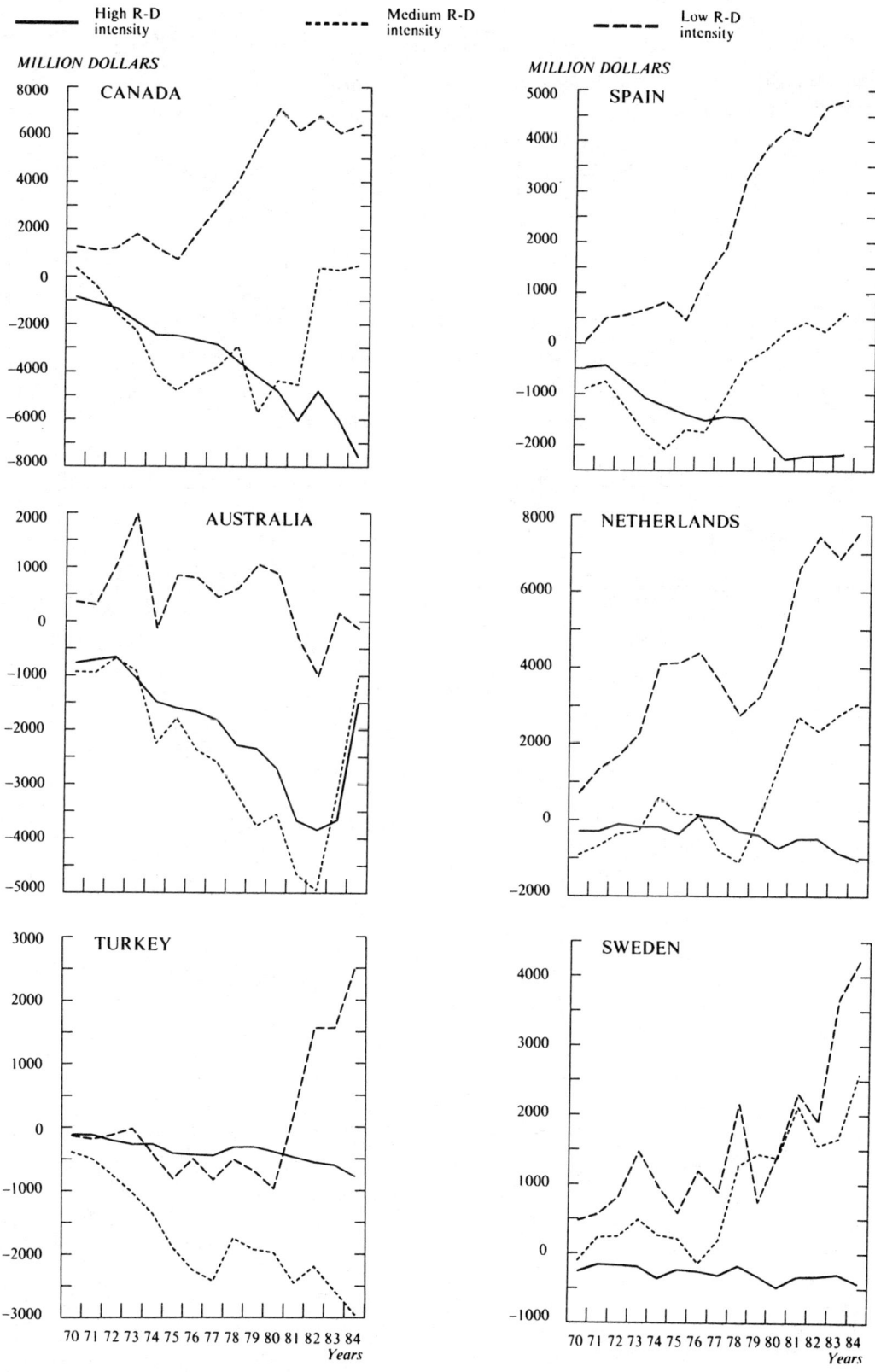

Graph 6 (cont'd)

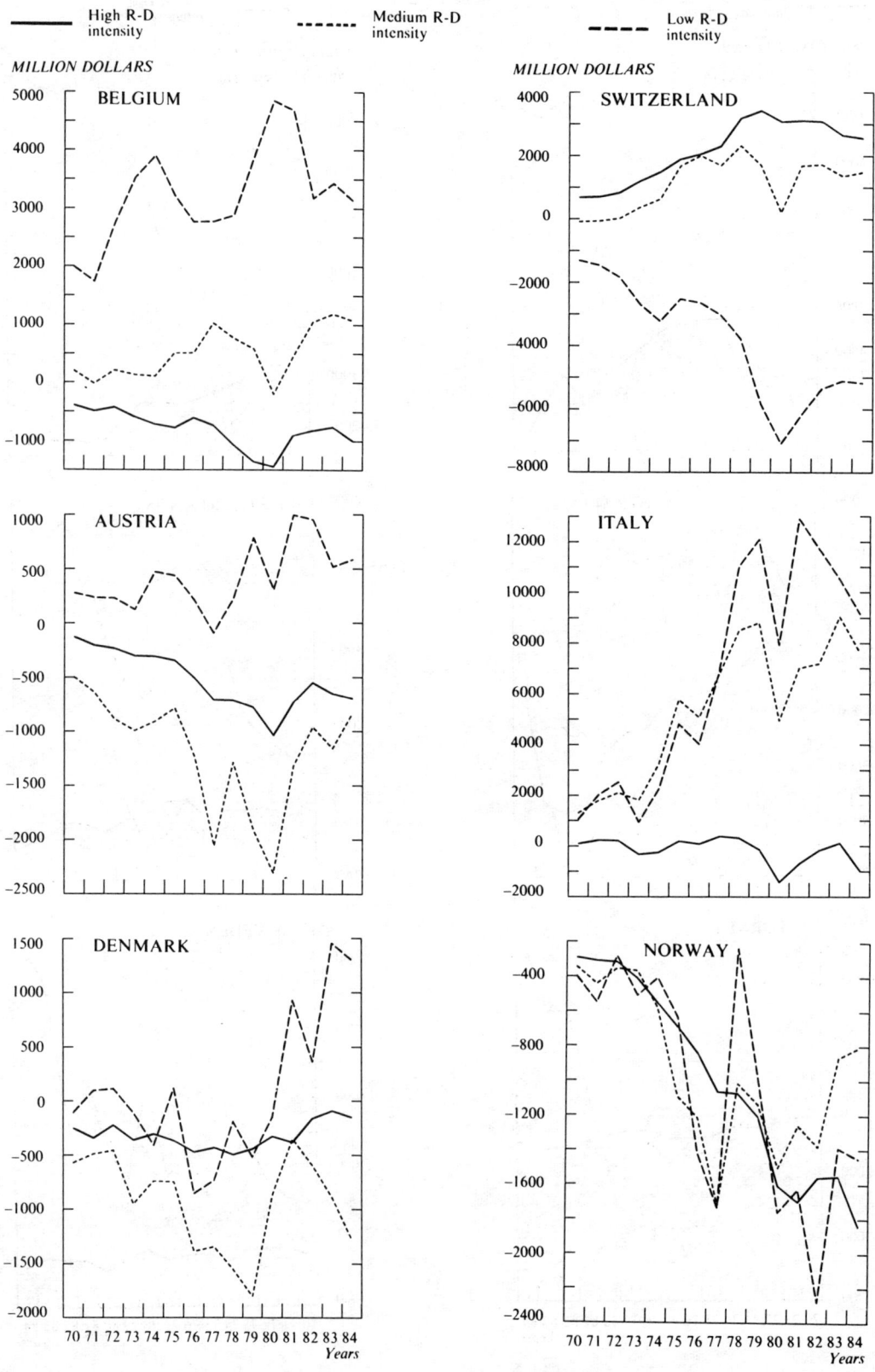

Graph 6 (cont'd)

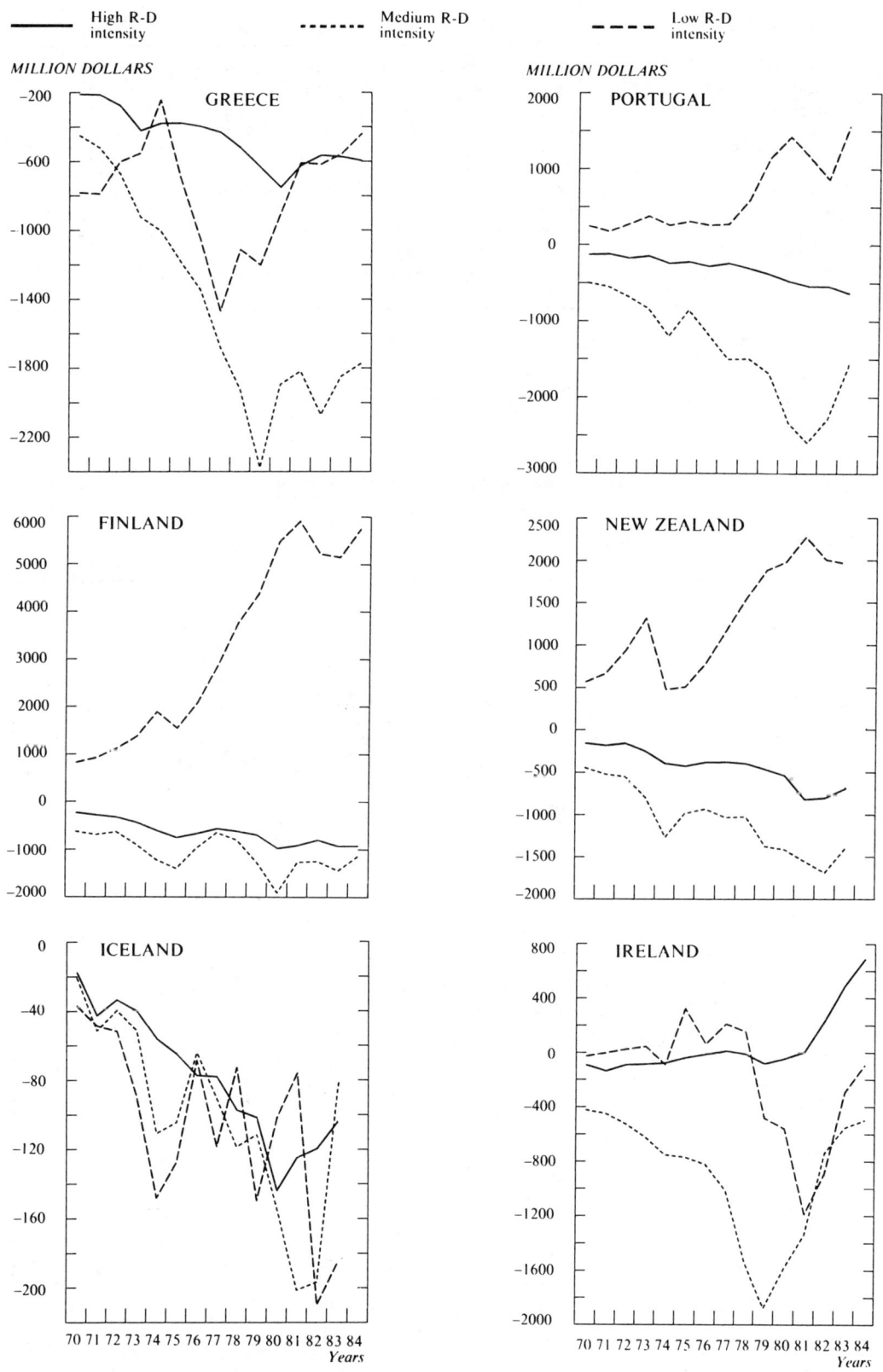

Table 35
Comparative unit labour costs in common currency terms
1970 = 100

	1975	1980	1981	1982	1983	1984[1]	1985[1]
United States	66.7	65.9	72.7	79.8	81.0	82.0	82.6
Japan	138.0	114.8	120.5	104.1	113.9	114.5	112.3
Germany	104.2	115.8	107.7	109.4	108.4	106.9	106.1
France	107.4	108.0	106.7	103.2	99.5	99.0	99.2
United Kingdom	98.5	141.6	145.2	135.8	126.1	127.6	129.7
Italy	105.7	92.6	91.2	93.0	101.9	106.7	110.4
Canada	96.3	85.0	91.9	101.7	105.4	104.3	104.3
Sweden	101.8	98.0	99.3	88.7	76.2	78.9	78.8
Netherlands	112.8	102.7	94.7	97.2	95.3	93.0	89.8
Belgium	105.0	101.5	92.8	78.7	74.0	71.5	71.9

1. Forecasts.

Graph 7
Relative unit labour costs in manufacturing
In common currency terms

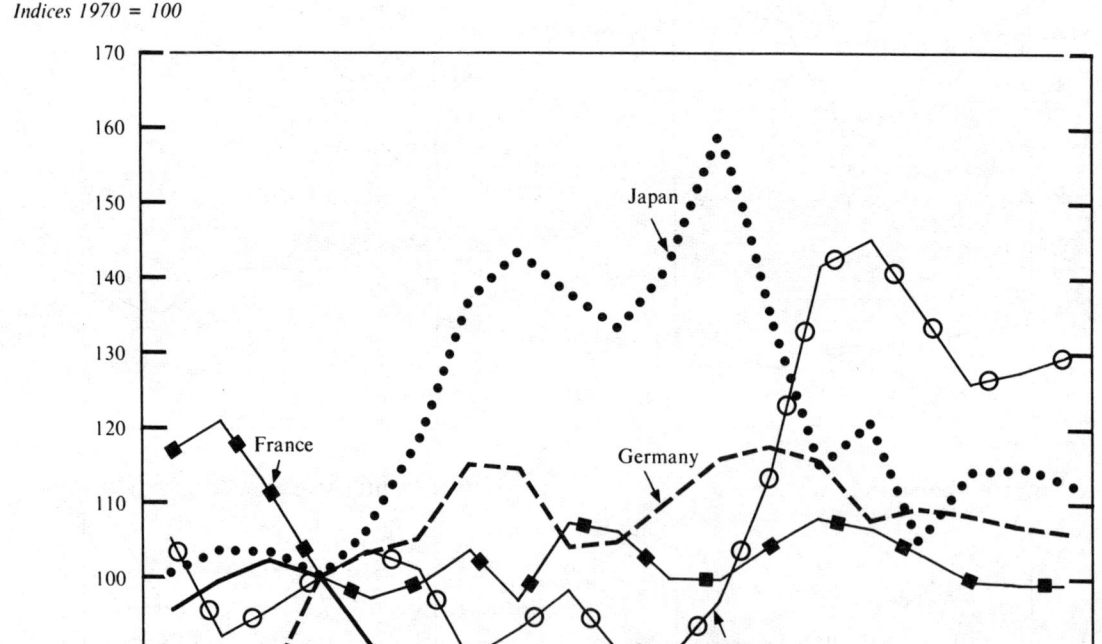

Source: OECD/STIIU, November 1985.

Table 36
Relative export prices of manufacturing industry
In common currency terms, 1970 = 100

	1975	1980	1981	1982	1983	1984[1]	1985[1]
United States	87.8	85.7	100.4	110.0	110.5	111.5	111.3
Japan	106.9	101.5	107.0	99.4	99.7	100.6	98.4
Germany	111.0	108.3	100.6	102.1	103.1	100.0	98.5
France	102.5	100.9	96.1	94.1	93.8	96.3	98.1
United Kingdom	93.7	118.8	114.4	106.8	105.0	104.0	106.1
Italy	94.9	95.3	94.4	94.5	93.4	95.5	97.2
Canada	87.2	80.9	81.1	79.7	82.9	82.3	82.4
Sweden	110.2	108.2	105.5	98.9	93.5	94.3	94.8
Netherlands	99.9	96.3	93.3	94.1	91.4	89.4	88.0
Belgium	99.5	105.0	97.6	92.8	91.2	91.2	92.0

1. Forecasts.

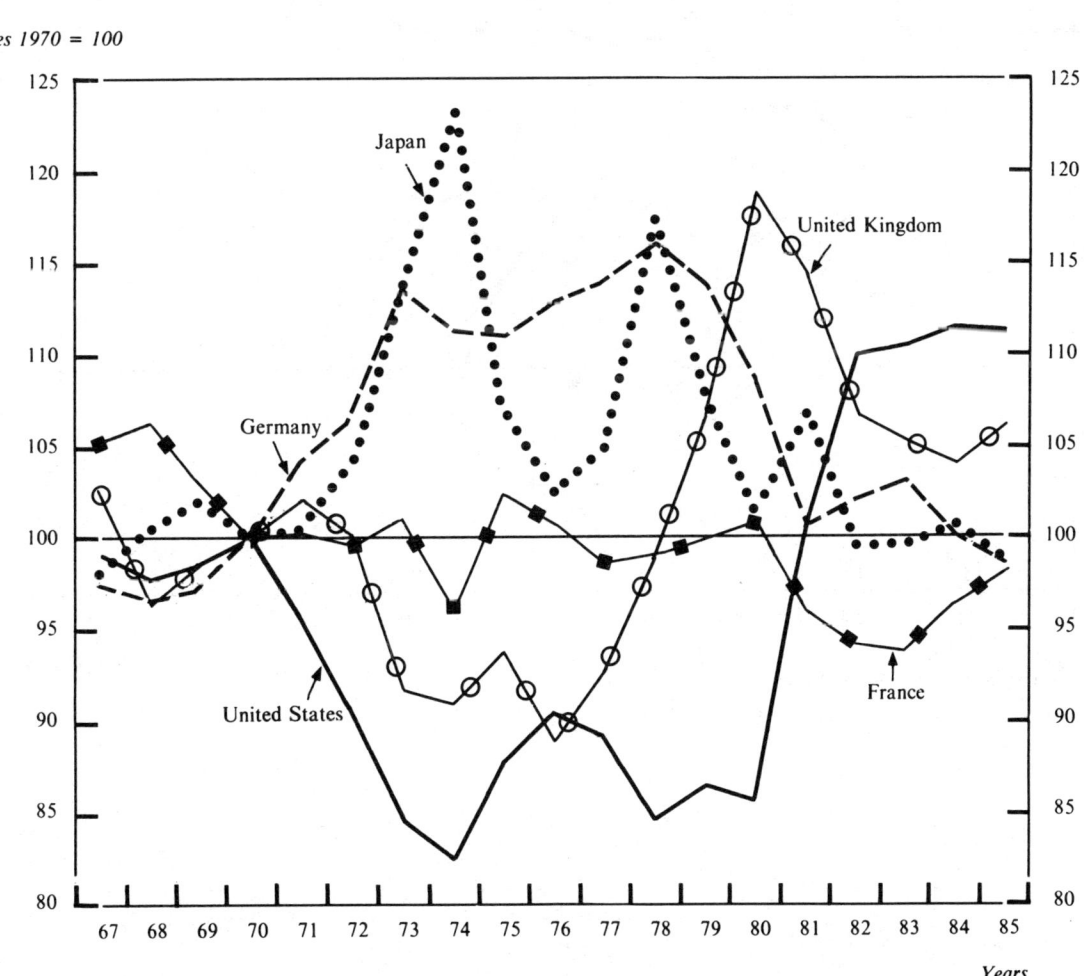

Graph 8
Relative export prices of manufacturing industry

Source: OECD/STIIU Data Bank, November 1985.

Graph 9
Elasticities of the high R&D industries
1970-1980

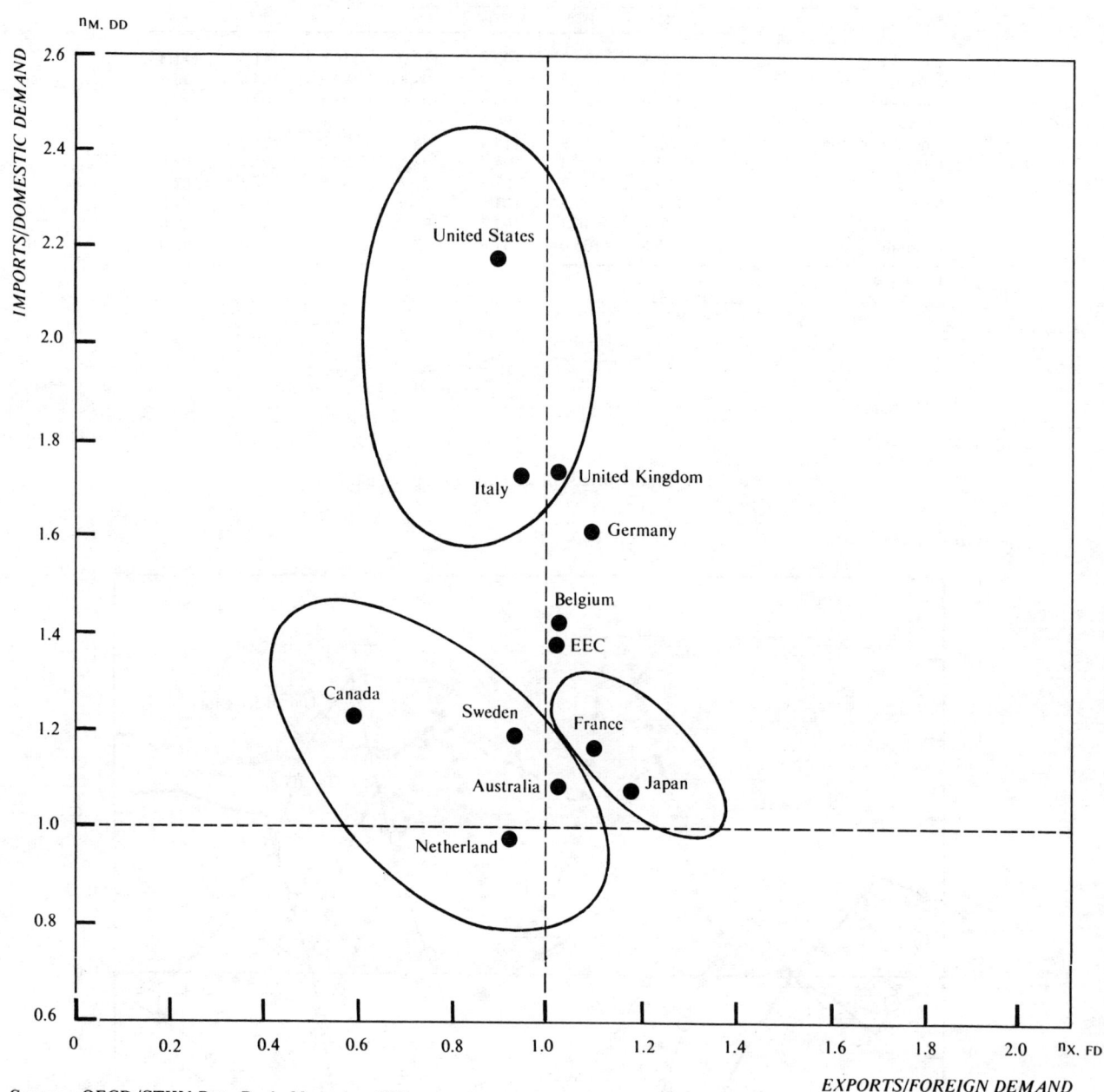

Source: OECD/STIIU Data Bank, November 1985.

Graph 10
Elasticities of the medium R&D industries
1970-1980

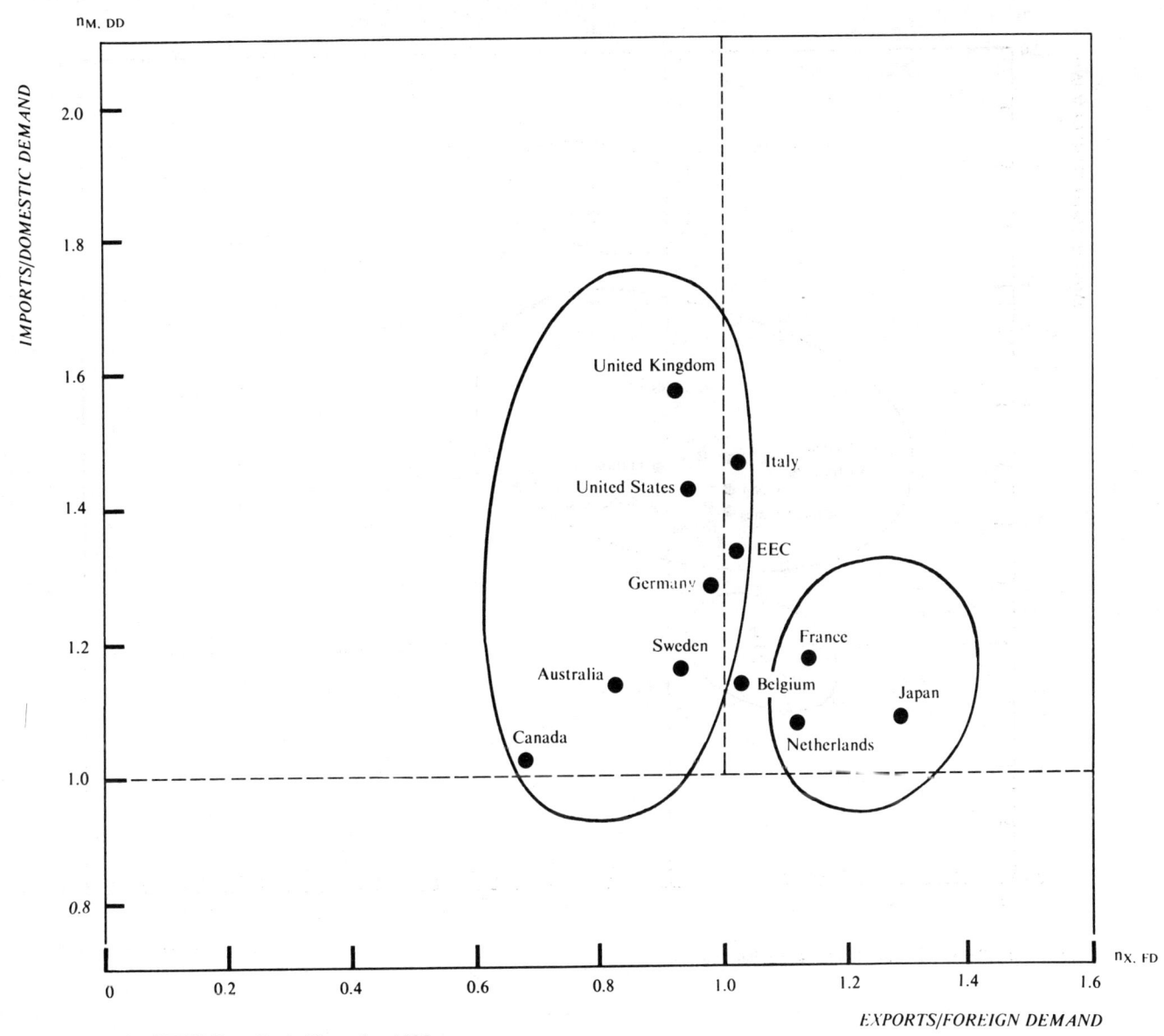

Source: OECD/STIIU Data Bank, November 1985.

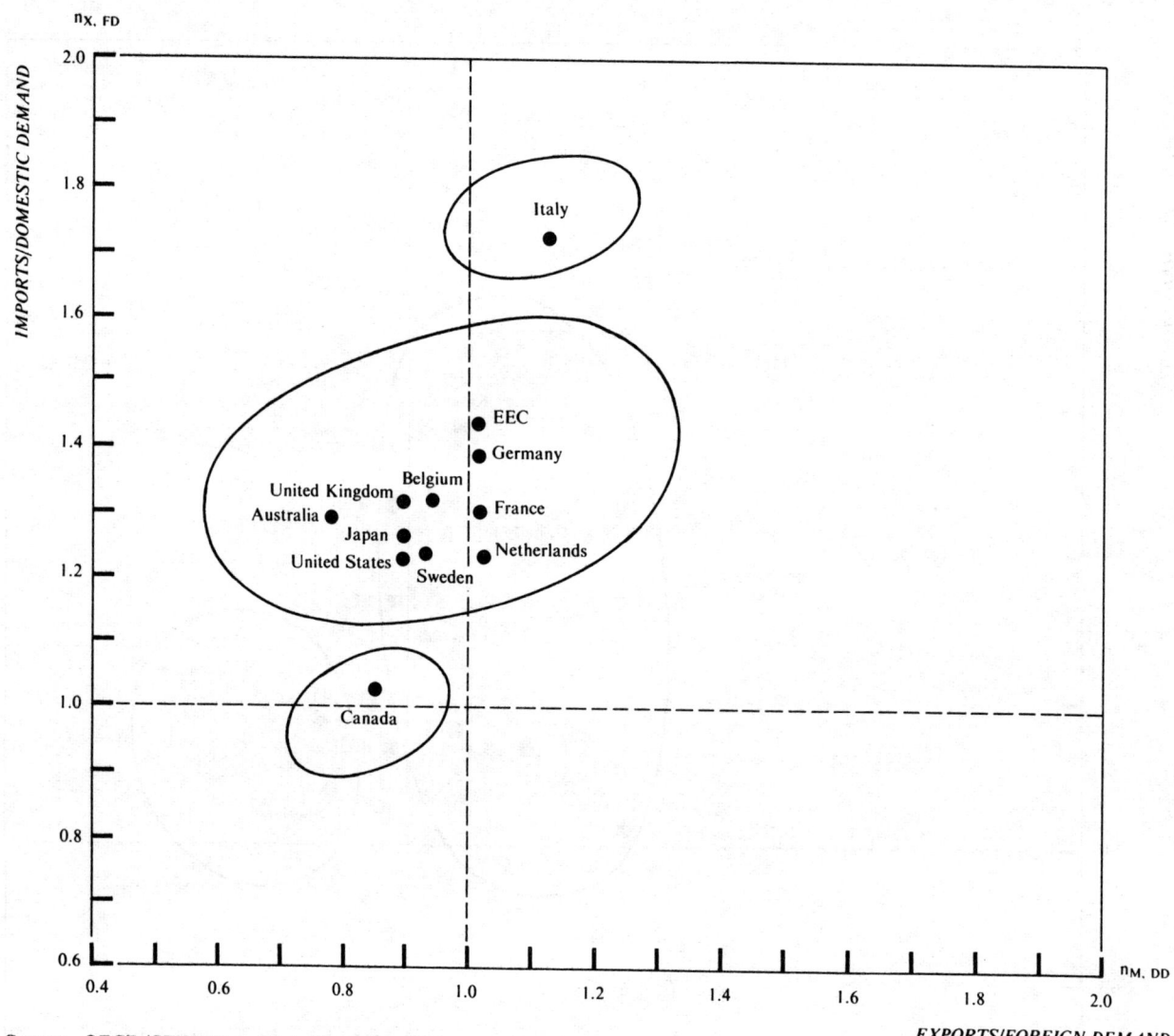

Graph 11
Elasticities of the low R&D industries
1970-1980

Source: OECD/STIIU Data Bank, November 1985.

Table 37

Elasticities of exports with respect to foreign demand 1970-1980

	United States	Japan	Germany	France	United Kingdom	Italy	Canada	Australia	Netherlands	Sweden	Belgium	EEC
High intensity												
Aerospace	0.87	0.51	1.92	1.16	1.27	1.12	0.29	1.82	0.95	. .	0.62	1.29
Computers	1.02	1.14	0.84	1.01	1.10	0.91	1.01	1.87	0.99	0.88	1.04	0.96
Electronics, components	0.92	1.12	0.98	1.12	0.95	0.82	0.64	0.83	0.89	0.93	0.98	0.96
Drugs and medecine	1.02	0.93	0.93	1.12	0.96	0.90	0.63	0.71	0.89	1.37	1.25	0.99
Instruments	0.91	1.25	0.90	1.03	0.94	0.91	0.72	1.15	0.98	1.01	1.35	0.95
Electrical machinery	0.95	1.19	0.99	1.08	0.88	1.07	0.58	0.70	0.85	0.92	0.90	0.98
Medium intensity												
Motor vehicles	0.87	1.52	0.99	1.10	0.73	0.96	0.66	1.10	1.15	0.97	1.06	0.98
Chemicals	0.96	0.90	0.96	1.13	0.99	0.96	0.98	1.15	1.07	0.92	1.05	1.02
Other manufacturing industries	0.88	0.75	0.82	1.12	1.10	1.18	0.91	0.86	1.00	0.88	1.19	1.07
Non-electrical machinery	0.95	1.36	0.94	1.00	0.88	1.01	0.87	0.54	1.04	0.93	0.98	0.63
Rubber, plastics	0.96	0.80	1.11	1.04	0.74	1.01	1.18	0.19	0.99	0.94	1.11	1.04
Non-ferrous metals	1.12	1.40	1.25	1.27	0.99	1.21	0.51	0.87	1.34	1.01	0.87	1.11
Low intensity												
Stone, clay, glass	0.91	1.00	0.97	1.05	0.90	1.23	0.99	0.82	1.01	0.98	0.78	1.01
Food, drink	0.95	0.52	1.38	1.06	1.01	1.07	0.70	0.73	0.99	0.76	1.11	1.09
Shipbuilding	0.89	1.09	0.91	1.29	1.04	0.80	0.74	1.16	0.80	0.76	0.50	0.97
Petroleum refineries	0.62	0.97	0.80	1.12	1.02	0.88	1.25	1.11	1.09	1.34	1.22	1.02
Ferrous metals	0.77	1.13	1.04	1.03	0.69	1.34	0.95	1.20	0.91	0.90	0.77	0.97
Fabricated metal products	0.95	1.01	0.98	1.12	0.82	1.23	0.83	0.62	0.96	0.92	1.02	1.02
Paper, printing	0.94	1.00	1.27	1.17	1.04	1.16	0.89	0.82	1.02	0.92	0.97	1.13
Woodwork, furniture	1.03	0.29	1.13	1.09	1.23	1.45	0.88	1.70	1.04	0.83	0.89	1.16
Textiles, footwear, leather	1.26	0.58	1.10	0.96	0.97	1.22	0.70	1.14	0.83	0.94	0.88	1.04
Total manufacturing industries	1.01	1.09	1.01	1.07	0.94	1.05	0.74	0.81	0.99	0.89	0.97	1.01

Source: OECD, STIIU, November 1985.

Table 38
**Elasticities of imports with respect to foreign demand
1970-1980**

	United States	Japan	Germany	France	United Kingdom	Italy	Canada	Australia	Netherlands	Sweden	Belgium	EEC
High intensity												
Aerospace	3.20	0.80	1.63	0.80	1.84	1.33	0.83	1.09	0.78	0.59	0.81	1.33
Computers	1.38	0.60	1.17	1.07	1.39	1.10	1.25	0.80	0.88	1.66	0.98	1.28
Electronics, components	2.18	1.59	1.51	1.21	1.77	1.33	1.36	1.43	1.03	0.98	1.22	1.42
Drugs and medecine	1.99	1.07	1.23	1.01	1.23	1.64	1.30	0.68	1.01	1.07	1.14	1.20
Instruments	1.69	1.18	1.34	1.27	1.40	1.51	1.11	1.00	0.51	1.09	1.38	1.30
Electrical machinery	2.27	1.26	1.74	1.42	1.72	1.32	1.26	1.28	1.06	1.23	1.17	1.48
Medium intensity												
Motor vehicles	1.35	1.33	1.15	1.20	2.04	1.33	1.00	1.27	0.98	0.99	1.14	1.29
Chemicals	1.29	1.16	1.33	1.25	1.43	1.42	1.20	1.15	1.06	1.22	1.19	1.33
Other manufacturing industries	1.82	0.90	1.42	1.42	1.56	..	1.48	1.20	1.10	1.18	0.47	1.56
Non-electrical machinery	1.91	0.92	1.36	1.07	1.48	1.78	1.11	1.23	1.04	1.23	1.08	1.28
Rubber, plastics	1.40	1.45	1.45	1.21	1.79	1.82	1.25	1.24	1.12	1.27	1.20	1.42
Non-ferrous metals	1.36	1.48	1.29	1.11	1.34	1.26	..	0.60	1.31	1.13	0.76	1.24
Low intensity												
Stone, clay, glass	1.57	1.27	1.54	1.12	1.59	1.90	1.15	1.09	1.06	1.34	1.33	1.45
Food, drink	1.11	1.06	1.15	1.30	0.99	1.35	1.32	1.36	1.26	1.14	1.30	1.21
Shipbuilding	0.89	2.13	0.93	1.11	1.65	1.24	0.92	1.04	0.93	1.07	0.76	1.17
Petroleum refineries	0.93	0.85	1.63	1.33	1.00	1.47	0.39	1.51	1.11	0.88	1.37	1.36
Ferrous metals	1.17	1.25	1.25	1.40	1.48	1.62	0.98	1.04	2.48	1.19	1.32	1.38
Fabricated metal products	1.48	1.17	1.50	1.45	2.92	2.67	1.10	1.10	1.19	1.35	1.38	1.66
Paper, printing	1.22	1.22	1.23	1.13	1.24	1.69	1.28	1.01	1.03	1.13	1.15	1.24
Woodwork, furniture	1.52	1.47	1.16	1.37	1.10	2.61	1.39	1.29	1.09	1.39	1.25	1.27
Textiles, footwear, leather	1.87	1.58	1.60	1.69	1.88	2.73	1.25	1.41	1.32	1.44	1.26	1.72
Total manufacturing industries	1.35	1.15	1.36	1.24	1.49		1.14	1.20	1.13	1.19	1.22	1.36

Source: OCDE, STIIU, November 1985.

Table 39
Shares of output of manufacturing industry in total output for eleven countries

	High R&D			Medium R&D			Low R&D		
	1970	1975	1980	1970	1975	1980	1970	1975	1980
United States	57.7	43.1	38.4	47.4	41.3	41.0	44.7	39.6	41.1
Japan	15.8	18.1	21.7	15.0	16.9	17.7	14.2	17.4	18.0
Germany	8.8	10.7	10.8	11.3	12.0	12.5	9.1	9.5	9.0
France	5.6	8.4	8.9	6.5	8.3	8.8	7.7	9.3	9.0
United Kingdom	7.2	6.8	7.8	7.4	6.9	6.6	8.0	7.3	7.4
Italy	4.6	4.6	4.9	4.7	4.9	4.5	5.5	4.7	4.7
Canada	2.1	2.2	1.8	2.8	3.0	2.7	3.9	4.0	3.6
Australia	0.7	1.0	0.88	1.2	1.3	1.3	1.4	1.6	1.4
Netherlands	1.3	2.1	2.2	0.9	1.4	1.5	1.8	2.3	2.3
Sweden	1.1	1.3	1.4	1.3	1.5	1.4	1.7	2.0	1.7
Belgium	0.65	0.93	0.84	1.2	1.7	1.5	1.5	1.7	1.5
Total	100.0	100.0	100.0	100.0	100.0	100.0	100.0	100.0	100.0
EEC	28.4	33.7	35.6	32.1	35.5	35.6	33.9	35.1	34.0

Source : OECD, STIIU, November 1985.

Graph 12
High R&D intensity industries

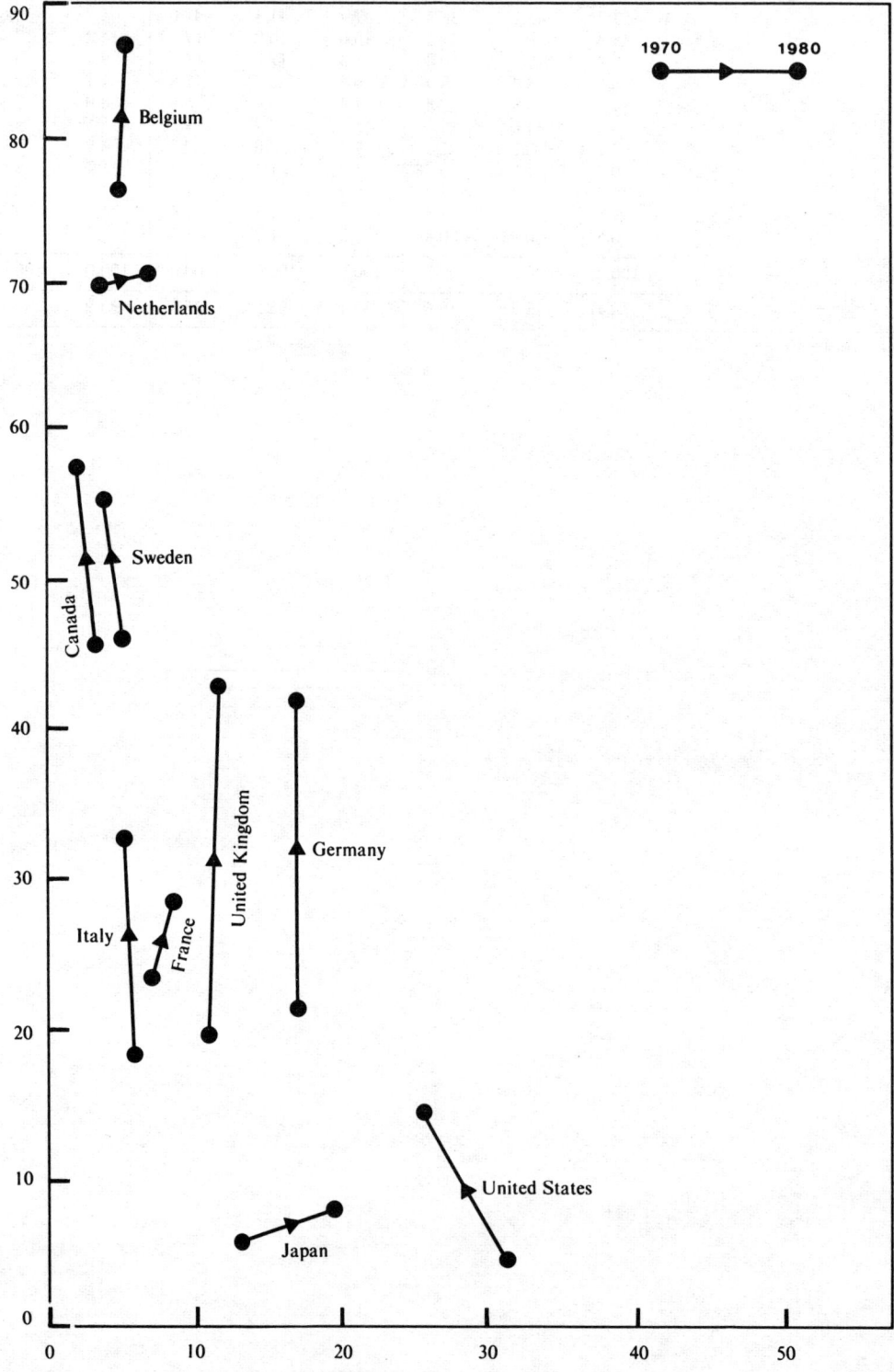

Source: OECD/STIIU Data Bank, November 1985.

Table 40
Share of manufacturing industry in OECD output (11 countries)

	United States		Japan		Germany		France		United Kingdom		Italy		Canada		Australia		Sweden		Netherlands		Belgium	
	1970	1980	1970	1980	1970	1980	1970	1980	1970	1980	1970	1980	1970	1980	1970	1980	1970	1980	1970	1980	1970	1980
High intensity																						
Aerospace	80.9	60.6	0.9	1.0	2.5	5.8	4.7	14.3	5.4	10.2	1.3	1.6	2.0	2.5	0.4	0.9	1.0	1.5	0.4	1.6	..	0.9
Computers	52.3	48.6	10.5	17.0	10.1	10.3	4.9	5.8	6.4	6.3	10.1	6.0	1.3	0.7	0.2	1.0	2.1	1.2	0.9	1.8	0.7	1.2
Electronics – components	48.4	33.3	21.7	28.8	7.4	9.0	4.3	8.1	8.4	7.5	2.9	4.6	1.8	1.4	0.6	0.6	1.0	1.8	2.2	3.3	0.8	1.2
Drugs and medecine	41.6	33.2	16.6	18.5	10.5	13.9	9.9	13.2	5.9	7.7	8.7	6.3	2.1	1.5	0.9	1.1	0.7	1.0	1.5	1.8	1.1	1.4
Instruments	51.9	44.1	18.9	26.0	12.3	13.0	4.1	4.3	6.8	7.2	2.3	1.4	2.0	1.5	0.3	0.5	0.6	0.7	0.2	0.8	0.08	0.03
Electrical machinery	42.7	31.2	17.8	23.0	11.3	12.4	6.6	9.2	7.7	8.1	6.0	7.4	2.8	2.5	1.1	1.0	1.3	1.5	1.5	2.3	0.8	0.9
Medium intensity																						
Motor vehicles	49.8	40.4	13.9	19.6	10.8	13.5	6.0	8.2	6.9	4.6	3.6	4.3	4.5	4.2	1.3	1.0	1.2	1.7	0.2	0.5	1.3	1.6
Chemicals	45.8	43.6	14.1	14.3	12.5	13.0	6.0	7.8	7.8	7.3	5.3	4.3	2.5	2.3	1.1	1.2	0.9	0.8	2.0	3.2	1.5	1.9
Other manufacturing industries	54.3	40.0	16.0	25.5	8.1	8.3	6.3	9.7	5.1	5.9	4.5	4.5	3.3	2.5	0.9	1.0	0.6	0.7	0.3	0.7	0.1	0.2
Non-electrical machinery	46.1	38.1	15.7	17.5	12.3	14.3	5.8	9.8	7.2	7.9	5.5	5.1	1.7	1.9	0.8	0.7	1.7	1.9	0.6	1.0	1.0	1.2
Rubber, plastics	45.3	39.3	17.1	23.8	7.3	7.2	6.9	10.0	9.2	6.9	6.0	5.0	2.8	2.3	1.5	1.4	1.1	0.9	1.1	1.3	1.2	1.4
Non-ferrous metals	47.7	46.4	15.2	12.3	11.2	9.9	7.2	9.9	7.2	6.9	3.0	3.0	2.1	1.7	1.9	4.6	1.7	1.6	0.9	1.5	1.4	1.8
Low intensity																						
Stone, clay, glass	38.7	31.0	16.5	22.2	11.5	11.0	6.2	10.7	7.6	7.5	9.2	7.2	2.9	2.7	1.6	1.8	1.6	1.3	1.4	1.8	2.3	2.4
Food, drink	46.7	40.0	10.6	16.0	7.2	8.4	9.5	10.6	9.2	8.7	4.8	4.6	4.2	3.4	1.9	1.9	1.5	1.3	2.7	3.3	1.2	1.4
Shipbuilding	34.1	39.6	28.6	22.1	8.1	6.4	5.9	7.2	7.0	6.3	3.5	3.8	2.2	2.1	1.1	1.4	4.9	3.8	3.0	4.3	1.1	2.6
Petroleum refineries	49.1	51.2	10.1	15.1	8.5	3.7	12.3	9.9	7.7	7.4	3.6	3.5	3.5	3.0	0.4	0.2	0.6	1.4	2.2	3.1	1.3	1.1
Ferrous metals	34.5	34.5	20.6	23.2	12.3	11.5	7.9	7.8	6.9	7.5	6.8	5.7	2.7	2.9	1.5	1.5	1.7	1.5	1.6	0.8	3.1	2.7
Fabricated metal products	45.4	42.6	11.7	20.7	11.2	13.0	4.1	5.6	8.8	5.6	3.9	3.2	3.3	3.3	1.3	1.6	1.7	1.7	1.2	1.5	0.9	0.9
Paper, printing	53.6	46.3	11.6	15.8	6.0	7.1	5.3	7.2	7.4	7.0	3.8	2.9	5.5	5.3	1.1	1.3	2.8	3.2	1.6	2.4	0.8	0.9
Wood, cork, furniture	40.7	34.2	20.1	21.8	8.5	12.3	6.2	7.6	6.0	5.6	4.9	3.3	5.8	6.2	1.6	1.8	3.5	3.8	1.0	1.4	1.3	1.6
Textiles, footwear, leather	44.0	37.2	12.5	17.6	11.0	11.6	8.2	10.1	7.8	7.1	8.1	8.3	3.2	3.3	1.1	1.1	0.8	0.6	1.4	1.1	1.4	1.5

Source: OECD, STIIU, November 1985.

OECD SALES AGENTS
DÉPOSITAIRES DES PUBLICATIONS DE L'OCDE

ARGENTINA - ARGENTINE
Carlos Hirsch S.R.L.,
Florida 165, 4º Piso,
(Galeria Guemes) 1333 Buenos Aires
Tel. 33.1787.2391 y 30.7122

AUSTRALIA-AUSTRALIE
D.A. Book (Aust.) Pty. Ltd.
11-13 Station Street (P.O. Box 163)
Mitcham, Vic. 3132   Tel. (03) 873 4411

AUSTRIA - AUTRICHE
OECD Publications and Information Centre,
4 Simrockstrasse,
5300 Bonn (Germany)   Tel. (0228) 21.60.45
Local Agent:
Gerold & Co., Graben 31, Wien 1 Tel. 52.22.35

BELGIUM - BELGIQUE
Jean de Lannoy, Service Publications OCDE,
avenue du Roi 202
B-1060 Bruxelles   Tel. 02/538.51.69

CANADA
Renouf Publishing Company Limited/
Éditions Renouf Limitée Head Office/
Siège social – Store/Magasin :
61, rue Sparks Street,
Ottawa, Ontario KIP 5A6
Tel. (613)238-8985. 1-800-267-4164
Store/Magasin : 211, rue Yonge Street,
Toronto, Ontario M5B 1M4.
Tel. (416)363-3171
Regional Sales Office/
Bureau des Ventes régional /
7575 Trans-Canada Hwy., Suite 305,
Saint-Laurent, Quebec H4T 1V6
Tel. (514)335-9274

DENMARK - DANEMARK
Munksgaard Export and Subscription Service
35, Nørre Søgade, DK-1370 København K
Tel. +45.1.12.85.70

FINLAND - FINLANDE
Akateeminen Kirjakauppa,
Keskuskatu 1, 00100 Helsinki 10 Tel. 0.12141

FRANCE
OCDE/OECD
Mail Orders/Commandes par correspondance :
2, rue André-Pascal,
75775 Paris Cedex 16
Tel. (1) 45.24.82.00
Bookshop/Librairie : 33, rue Octave-Feuillet
75016 Paris
Tel. (1) 45.24.81.67 or/ou (1) 45.24.81.81
Principal correspondant :
Librairie de l'Université,
13602 Aix-en-Provence   Tel. 42.26.18.08

GERMANY - ALLEMAGNE
OECD Publications and Information Centre,
4 Simrockstrasse,
5300 Bonn   Tel. (0228) 21.60.45

GREECE - GRÈCE
Librairie Kauffmann,
28 rue du Stade, Athens 132   Tel. 322.21.60

HONG KONG
Government Information Services,
Publications (Sales) Office,
Beaconsfield House, 4/F.,
Queen's Road Central

ICELAND - ISLANDE
Snæbjörn Jónsson & Co., h.f.,
Hafnarstræti 4 & 9,
P.O.B. 1131 – Reykjavik
Tel. 13133/14281/11936

INDIA - INDE
Oxford Book and Stationery Co.,
Scindia House, New Delhi 1   Tel. 45896
17 Park St., Calcutta 700016   Tel. 240832

INDONESIA - INDONESIE
Pdin Lipi, P.O. Box 3065/JKT.Jakarta
Tel. 583467

IRELAND - IRLANDE
TDC Publishers – Library Suppliers
12 North Frederick Street, Dublin 1
Tel. 744835-749677

ITALY - ITALIE
Libreria Commissionaria Sansoni,
Via Lamarmora 45, 50121 Firenze
Tel. 579751/584468
Via Bartolini 29, 20155 Milano   Tel. 365083
Sub-depositari :
Ugo Tassi, Via A. Farnese 28,
00192 Roma   Tel. 310590
Editrice e Libreria Herder,
Piazza Montecitorio 120, 00186 Roma
Tel. 6794628
Agenzia Libraria Pegaso,
Via de Romita 5, 70121 Bari
Tel. 540.105/540.195
Agenzia Libraria Pegaso, Via S.Anna dei
Lombardi 16, 80134 Napoli.   Tel. 314180
Libreria Hœpli,
Via Hœpli 5, 20121 Milano   Tel. 865446
Libreria Scientifica
Dott. Lucio de Biasio "Aeiou"
Via Meravigli 16, 20123 Milano Tel. 807679
Libreria Zanichelli, Piazza Galvani 1/A,
40124 Bologna   Tel. 237389
Libreria Lattes,
Via Garibaldi 3, 10122 Torino   Tel. 519274
La diffusione delle edizioni OCSE è inoltre
assicurata dalle migliori librerie nelle città più
importanti.

JAPAN - JAPON
OECD Publications and Information Centre,
Landic Akasaka Bldg., 2-3-4 Akasaka,
Minato-ku, Tokyo 107   Tel. 586.2016

KOREA - CORÉE
Pan Korea Book Corporation
P.O.Box No. 101 Kwangwhamun, Seoul
Tel. 72.7369

LEBANON - LIBAN
Documenta Scientifica/Redico,
Edison Building, Bliss St.,
P.O.B. 5641, Beirut   Tel. 354429-344425

MALAYSIA - MALAISIE
University of Malaya Co-operative Bookshop
Ltd.,
P.O.Box 1127, Jalan Pantai Baru,
Kuala Lumpur   Tel. 577701/577072

NETHERLANDS - PAYS-BAS
Staatsuitgeverij Verzendboekhandel
Chr. Plantijnstraat, 1 Postbus 20014
2500 EA S-Gravenhage   Tel. 070-789911
Voor bestellingen: Tel. 070-789208

NEW ZEALAND - NOUVELLE-ZÉLANDE
Government Printing Office Bookshops:
Auckland: Retail Bookshop, 25 Rutland Street,
Mail Orders, 85 Beach Road
Private Bag C.P.O.
Hamilton: Retail: Ward Street,
Mail Orders, P.O. Box 857
Wellington: Retail, Mulgrave Street, (Head
Office)
Cubacade World Trade Centre,
Mail Orders, Private Bag
Christchurch: Retail, 159 Hereford Street,
Mail Orders, Private Bag
Dunedin: Retail, Princes Street,
Mail Orders, P.O. Box 1104

NORWAY - NORVÈGE
Tanum-Karl Johan a.s
P.O. Box 1177 Sentrum, 0107 Oslo 1
Tel. (02) 801260

PAKISTAN
Mirza Book Agency
65 Shahrah Quaid-E-Azam, Lahore 3 Tel. 66839

PORTUGAL
Livraria Portugal,
Rua do Carmo 70-74, 1117 Lisboa Codex.
Tel. 360582/3

SINGAPORE - SINGAPOUR
Information Publications Pte Ltd
Pei-Fu Industrial Building,
24 New Industrial Road No. 02-06
Singapore 1953   Tel. 2831786, 2831798

SPAIN - ESPAGNE
Mundi-Prensa Libros, S.A.,
Castelló 37, Apartado 1223, Madrid-28001
Tel. 431.33.99
Libreria Bosch, Ronda Universidad 11,
Barcelona 7   Tel. 317.53.08/317.53.58

SWEDEN - SUÈDE
AB CE Fritzes Kungl. Hovbokhandel,
Box 16356, S 103 27 STH,
Regeringsgatan 12,
DS Stockholm   Tel. (08) 23.89.00
Subscription Agency/Abonnements:
Wennergren-Williams AB,
Box 30004, S104 25 Stockholm. Tel. 08/54.12.00

SWITZERLAND - SUISSE
OECD Publications and Information Centre,
4 Simrockstrasse,
5300 Bonn (Germany)   Tel. (0228) 21.60.45
Local Agent:
Librairie Payot,
6 rue Grenus, 1211 Genève 11
Tel. (022) 31.89.50

TAIWAN - FORMOSE
Good Faith Worldwide Int'l Co., Ltd.
9th floor, No. 118, Sec.2
Chung Hsiao E. Road
Taipei   Tel. 391.7396/391.7397

THAILAND - THAILANDE
Suksit Siam Co., Ltd.,
1715 Rama IV Rd.,
Samyam Bangkok 5   Tel. 2511630

TURKEY - TURQUIE
Kültur Yayinlari Is-Türk Ltd. Sti.
Atatürk Bulvari No: 191/Kat. 21
Kavaklidere/Ankara   Tel. 17.02.66
Dolmabahce Cad. No: 29
Besiktas/Istanbul   Tel. 60.71.88

UNITED KINGDOM - ROYAUME UNI
H.M. Stationery Office,
Postal orders only:
P.O.B. 276, London SW8 5DT
Telephone orders: (01) 622.3316, or
Personal callers:
49 High Holborn, London WC1V 6HB
Branches at: Belfast, Birmingham,
Bristol, Edinburgh, Manchester

UNITED STATES - ÉTATS-UNIS
OECD Publications and Information Centre,
Suite 1207, 1750 Pennsylvania Ave., N.W.,
Washington, D.C. 20006 - 4582
Tel. (202) 724.1857

VENEZUELA
Libreria del Este,
Avda F. Miranda 52, Aptdo. 60337,
Edificio Galipan, Caracas 106
Tel. 32.23.01/33.26.04/31.58.38

YUGOSLAVIA - YOUGOSLAVIE
Jugoslovenska Knjiga, Knez Mihajlova 2,
P.O.B. 36, Beograd   Tel. 621.992

Orders and inquiries from countries where Sales
Agents have not yet been appointed should be sent
to:
OECD, Publications Service, Sales and
Distribution Division, 2, rue André-Pascal, 75775
PARIS CEDEX 16.

Les commandes provenant de pays où l'OCDE n'a
pas encore désigné de dépositaire peuvent être
adressées à :
OCDE, Service des Publications. Division des
Ventes et Distribution. 2. rue André-Pascal. 75775
PARIS CEDEX 16.

69482-03-1986

OECD PUBLICATIONS, 2, rue André-Pascal, 75775 PARIS CEDEX 16 - No. 43479 1986
PRINTED IN FRANCE
(92 86 02 1) ISBN 92-64-12809-3